Palgrave Studies in Natural Resource Management

Series Editor
Justin Taberham
London, UK

This series is dedicated to the rapidly growing field of Natural Resource Management (NRM). It aims to bring together academics and professionals from across the sector to debate the future of NRM on a global scale. Contributions from applied, interdisciplinary and cross-sectoral approaches are welcome, including aquatic ecology, natural resources planning and climate change impacts to endangered species, forestry or policy and regulation. The series focuses on the management aspects of NRM, including global approaches and principles, good and less good practice, case study material and cutting edge work in the area.

More information about this series at
http://www.palgrave.com/gp/series/15182

Stephen Muddiman

Ecosystem Services

Economics and Policy

Stephen Muddiman
Harwood Biology
Great Harwood, UK

Palgrave Studies in Natural Resource Management
ISBN 978-3-030-13818-9   ISBN 978-3-030-13819-6   (eBook)
https://doi.org/10.1007/978-3-030-13819-6

Library of Congress Control Number: 2019931930

Cover credit: Peacock Graphics/Alamy Stock Photo

This Palgrave Macmillan imprint is published by the registered company Springer Nature Switzerland AG
The registered company address is: Gewerbestrasse 11, 6330 Cham, Switzerland

Contents

List of Figures

Parallels and Function

Valuing Ecosystems

List of Tables

Introduction

The Importance of Knowledge

There can be little doubt that the collective advancement of knowledge and understanding is a key driving force in the development of modern civilisations. Major advancements in agricultural, industrial, scientific and technological knowledge have been the main hallmarks of advanced civilisations throughout history.

In a perfect world, all of the foundations of civilisations would be logical and rational. Development would be based upon the solid foundation of accurate, verifiable and objective truth. It is, however, inherent in the development of greater knowledge that certain facts can be interpreted in a variety of ways to provide a particular world view. It is only by the addition of further knowledge that this view can be either strengthened, modified or superseded by a new set of ideas. For example, it has long been known that growing the same crop on an area of land for many years leads to a gradual decline in yield. The concept of leaving an area of land fallow in order for it to recover its fertility has been common practice since biblical times. The precise reasons for this practice in terms of nutrient cycles and the ongoing accumulation

© The Author(s) 2019
S. Muddiman, *Ecosystem Services,*
Palgrave Studies in Natural Resource Management,
https://doi.org/10.1007/978-3-030-13819-6_1

of pests over time were not understood fully by ancient farming communities. In this case, the accumulation of subsequent knowledge has not only served to strengthen the rationale for leaving land fallow, but also led to the development of crop rotation systems, which further enhanced both yields and agricultural sustainability.

So, in practical terms, there is no such thing as the perfect system, we are always constrained by the limits of our knowledge and also by the models we use to order and rationalise the facts we have to hand at any given point in time. Even today, there are many grey areas where 'truth' can be considered to be a matter of faith, opinion or pragmatism. For example, the concept of democracy as a means of government is generally seen as the only way in which a fair and free society can be established, but the truth of it is rule by majority opinion is often considered to be the 'least worst' of the available systems. As Winston Churchill noted in 1947 'it has been said that democracy is the worst form of Government except for all those other forms that have been tried from time to time....'. It is possible that the future will offer alternative approaches to the operation of governance which will improve upon the existing system. Democracy, like everything else, is a work in progress.

This book is an investigation into two particular 'works in progress': how we view, value and assess the natural environment; and how individuals, communities and nations interact with each other in terms of exchange of resources, products and skills. These elements both come together in the concept of Ecosystem Services which is an attempt to incorporate the natural environment into the sphere of human commercial activity. This idea is relatively recent in origin, and has grown out of the increasing awareness and concern about the potential conflicts between the growth of modern civilization and its effects on the elements of the natural environment upon which humanity depends. It is borne out of a fear that as we grow and develop as a species we run the risk of actually severing those vital lifelines which sustain us. As a result, the concept of Ecosystem Services offers the potential to link environmental degradation and loss, matters which are becoming increasingly prominent, with the sphere of economics and development. This linkage is achieved by adopting an anthropocentric view of the natural environment. It defines those services which are obtained from the natural operation of ecosystems (i.e. without the need for human labour or other resource input) and are of utility to humanity, as Ecosystem Services.

As a multidisciplinary concept, Ecosystem Services draws from both the current state of knowledge of ecology and economics and adopts the 'received wisdom' from both in terms of their respective mechanisms and modes of operation. In this way, the current framework of Ecosystem Services is based upon our understanding of ecology and economics and the models developed from the knowledge we hold at the present time. As such, it is important to gain an understanding of the extent of this knowledge and how close to a comprehensive framework each of these disciplines currently encompasses when discussing their role and validity in the derivative ideas of Ecosystem Services.

The truth is that both disciplines (ecology and economics) suffer from an acute lack of available data upon which to work from, so it is perfectly possible that the ecological and economic models which are currently being used as the basis for the development and implementation of Ecosystem Services practice may be erroneous to a greater or lesser degree.

Because of the knowledge deficit which, in many respects underlies this narrative, this is a book which explores the unknown. The following statement by US Secretary of Defence, Donald Rumsfeld is particularly pertinent in this respect:

> as we know, there are known knowns; there are things we know we know. We also know there are known unknowns; that is to say we know there are some things we do not know. But there are also unknown unknowns – the ones we don't know we don't know.

Leaving aside the 'unknown unknowns' for the time being, there is sufficient scope within the 'known unknown' category to offer a plausible case for reflection upon the possible implications of the way in which Ecosystem Services are being used to guide policy, and the way in which they have become embedded in the prevailing economic status quo.

In order to illustrate the knowledge deficit encountered when bringing together ecology and economics I will take two concepts, one from each discipline, which are frequently used as metrics in determining environmental and economic policy, respectively, and dig a little deeper into how much (or little) we can expect to understand by their application.

Biodiversity

The term Biodiversity, has been employed by ecologists and other environmental scientists to describe the variety of life encountered in the natural world. It is also probably the most elusive term to pin down in respect of a practical meaning. The most common definition is that used in the United Nations Convention on Biological Diversity:

> the variability among living organisms from all sources, including, *inter alia*, terrestrial, marine and other aquatic ecosystems and the ecological complexes of which they are part: this includes diversity within species, between species and of ecosystems.

Of course, in the provision of such a definition, the casual reader or policy maker would assume that we have some grasp on the amount of variability present in the natural world, certainly to the extent that it was possible to realistically commit to its protection. This is of particular relevance when the Convention requires signatories to prepare a 'National biodiversity Strategy'. In developing such a 'strategy' it would be realistic to assume that it would involve some mechanism to take account of 'the variability among living organisms from all sources' as encompassed in the above definition of Biodiversity. Unfortunately, this is far from the reality of the situation. We are in large part ignorant of the degree of variability present.

In ecology, the species is the basic unit from which other concepts such as habitats, communities and assemblages derive. They are all fundamentally composed of populations of different species. On that basis, it would be reasonable to assume that there was a requirement to at least have an approximate idea of the number of species in existence on the planet and how they are distributed amongst the various described habitats, communities, etc. This is of particular importance when we speak of conserving or enhancing biodiversity. Such actions must surely be taking account of those species which actually comprise the corpus of 'biodiversity' which is being conserved?

The truth is that we have managed to describe somewhere in the region of 1.6 million species to date [1], but estimates of how many species exist varies widely. One of the most statistically valid analyses comes from a study carried out in 2011 [2]. This research determined the presence of statistical correlations between higher taxonomic levels (Phylum, Class, Order, Family, Genus) and the number of species occurring within each. The researchers quantified these relationships enabling them to predict how many species there are likely to be. The figure derived using this method was 8.7 ($\pm$1.3) million species on earth and a further 2.2 ($\pm$0.18) million species in the oceans. The following figure, derived from this study, shows the likely number of organisms of selected groups, and the current state of knowledge in terms of percentages (Table 1).

The level of ignorance of biodiversity is not evenly distributed around the world. The majority of undescribed species are thought to occur at lower latitudes or in the oceans. Due to this pattern, a location such as the British Isles, with a long heritage of biological recording, a relatively depauperate fauna and flora and a high population density would be expected to be well known in terms of the biodiversity present. However, developments in DNA analysis show that this confidence is misplaced. One recent study on the slugs of Britain and Ireland, which

Table 1 Predicted Global Species Diversity

Group	Earth			Oceans		
	Described	Predicted	% known	Described	Predicted	% known
Animals	953,434	7,770,000	12.3	171,082	2,150,000	8.0
Plants	215,664	298,000	72.4	8600	16,600	51.8
Fungi	43,271	611,000	7.1	1097	5320	20.6

Source From data in Mora et al. [2]

are a well-studied fauna of economic importance [3] identified eight previously undetected species (which constitutes 22% of the fauna). This clearly illustrates the degree to which we are in the dark about the numbers of species in the world.

With such a lack of knowledge within such a well-defined concept as the 'species', the adoption of a definition of biodiversity which includes 'diversity within species' leads to a mind-boggling level of ignorance. This is compounded when the rest of the definition is taken into account, by including 'ecosystem diversity' one finds oneself charged with the responsibility to maintain every unique combination of life. This is clearly not the intention of the Convention, but by casting such a broad net in respect of the definition of biodiversity in order to satisfy every ecological concern it has meant that the implementation of the Convention is so open to interpretation and pragmatic compromise that almost anything can be considered acceptable.

When faced with the question 'how much biodiversity is there on the planet?' the only honest answer that can be given is quite simply: We have no idea.

Gross Domestic Product

In economics, the Gross Domestic Product (GDP) is defined as the monetary value of all the finished goods and services produced within a country in a specific time period. It is commonly used as an indicator of the economic health and standard of living of a country. The concept of GDP first came into use in 1937, developed by Russian economist Simon Kuznets. It was presented in a report to the US Congress when it was decided that there was a need to develop more meaningful national accounting in response to the Great Depression.

Unlike the concept of biodiversity, GDP has a well-formed definition, and an established methodology for collecting the necessary data. It also has the advantage of being expressed as a single numerical value. The difficulty with GDP is the lack of knowledge regarding the meaning and implications of this apparently concrete and definitive metric.

Growth in GDP is considered to be a sign that an economy is doing well. On the contrary, a shrinking GDP is considered to be a bad thing (the formal definition of a 'recession' is a shrinking GDP in two consecutive quarters). In turn, the readings of GDP are used by Governments to establish policies which are intended to promote growth (as measured by an increase in GDP).

As can be seen in Fig. 1, GDP on a global scale has increased almost exponentially since 1960. This is generally interpreted to mean that the world's economy has been constantly growing, with more goods and services being produced and economic well-being improving significantly.

Superficially, it appears that the measurement of GDP and policy decisions are a perfectly straightforward relationship, economic policies which enhance GDP are seen as fundamentally beneficial. This is how GDP is presented in general, being used to publicise how well or otherwise one's national economy is doing. However, a more in-depth study of the concept of GDP shows that it is far from a problem-free metric.

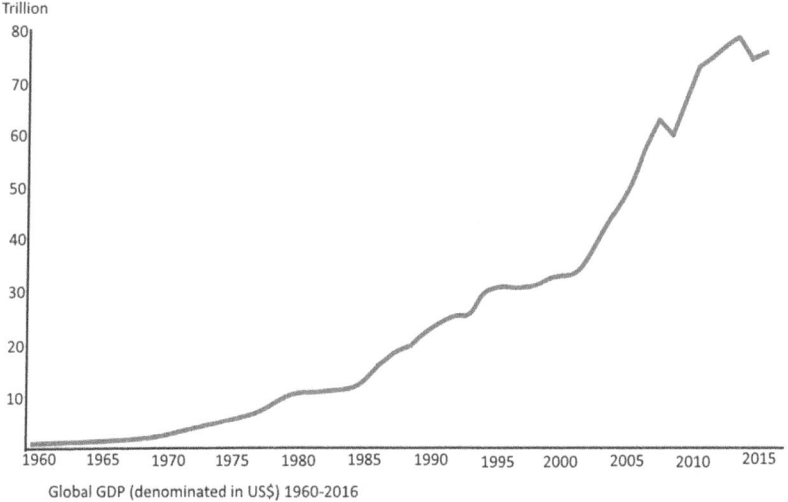

Fig. 1 Global GDP (*Source* World Bank national accounts data and OECD national accounts data files)

Table 2   Causes of GDP growth

Change	Cause	Implication
More production at the same prices	Production is increasing to meet increased demand	Lower unemployment rate and higher wages. This leads to even higher demand. Resulting in further increases in GDP, alongside increasing production prices due to rising labour costs (inflation)
Production is stable, but prices have increased	There has been an increase in the cost of production as a result of increases in the prices of raw materials or energy	GDP rises, but prices also rise due to rising materials costs (inflation)
More production at higher prices	There is both increased demand and shortage of supply	Increased demand and decreased supply leads to rapid price increases. GDP and inflation both increase at a rapid rate
Produced much more at lower prices	Overproduction, leading to falls in prices of excess goods	employment and wages drop, reducing demand, furthering overproduction—deflation
Produced less at much higher prices	Prices increase due to production costs, but demand is diminishing due to high unemployment and inflation	GDP rises slowly, below the desired level, yet inflation persists and unemployment remains high due to low production—'stagflation'

Source Various sources

The growth of GDP indicates that one of five possible changes have occurred within an economy. These are explained in Table 2.

As Table 2 shows, an increase in GDP isn't always good news. It isn't a clear signal of prosperity and can come about in several disparate ways. Interpretation of the figures is vital in order to determine how a rising GDP has occurred. This in turn leads to policy decisions regarding a range of economic activities. Central Banks may raise or lower interest rates, Governments may decide they can raise or lower taxes or borrow money for additional spending.

As some of the underlying reasons for increase in GDP show, it is clear that GDP is a measure of production which does not consider economic well-being. Growth in GDP can be observed in situations where unemployment is rising and prices are increasing. In no way, therefore, can this measure be considered to indicate the prosperity of a nation at the level of individuals. Rather it is something which can be used to look at a more generalised situation, and also to compare how one nation is doing in competition with others, in terms of the narrowly defined concept of productivity.

There is an additional factor in GDP which is generally overlooked by mainstream economics. That is the role of Central Banks, those institutions which are literally responsible for the manufacture of money. As a rule, the more money created by the central bank and the banking sector, the larger the monetary spending will be. This in turn means that the rate of growth of what is considered the 'real' economy will closely mirror rises in money supply. In fact, irrespective of how well an economy is doing, a Central Bank can stimulate the GDP figure by increasing the money supply. Of course, this is an illusory stimulus, as inflation is also stimulated by this action. It can be considered more of a recalibration for accounting purposes rather than an actual stimulus which benefits the population.

One significant element of a nation's economic life which is not reflected in the GDP figure is the amount of work carried out on one's own behalf. The degree to which tasks are carried out independently (such as cooking at home rather than using restaurants, or caring for children within a family rather than paying for childcare) do not count towards a nations GDP, even though the work has been carried out. The lack of a financial transaction makes this labour 'invisible'. In this way, the GDP framework has the unfortunate unintended consequence of giving the impression that it is not the activities of individuals that produce goods and services, but something else outside these activities called the 'economy'. However, in reality at no stage does this abstract 'economy' exist independently of the activities of individuals.

Paradoxically, undertaking work on one's own behalf can be seen as undermining the nation's economic success, as reflected in GDP figures.

The use of GDP as a widespread index does not consider the environmental costs associated with increased productivity. As such, it ignores environmental degradation through landtake, the accommodation of waste products and pollution. All of which have the effect of not only affecting the environment but the associated quality of life of the population who are exposed to such degradation.

In fact, some negative environmental effects such as discrete pollution incidents, can actually have a positive impact on a nations GDP as the costs of clean-up and other activities involve commercial transactions which are reflected in GDP calculations. For example, the increase in the GDP of Japan in 2011–2013 is correlated with the tsunami and subsequent breach of the Fukushima Nuclear Reactor, which took place in October 2011 (Fig. 2).

In the classic textbook, 'Economics' [4] GDP is considered to be analogous a satellite in space that can survey the weather across an entire continent. GDP enables policymakers to judge whether the

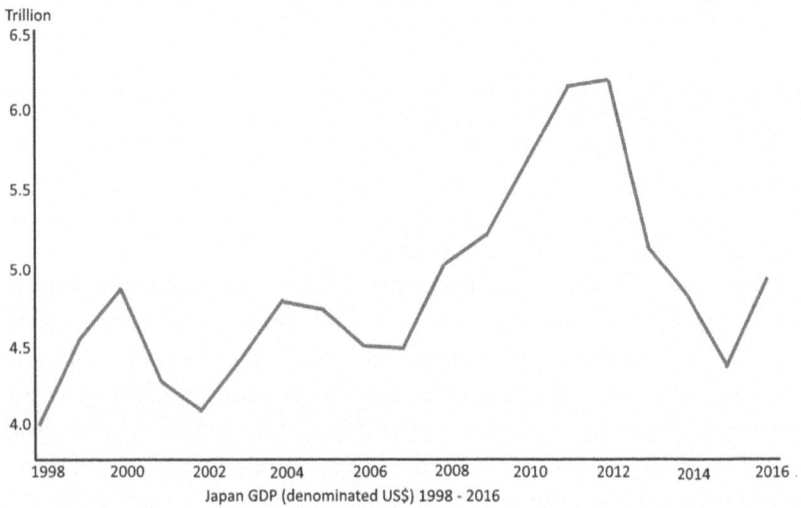

Fig. 2 Japan GDP (*Source* World Bank national accounts data and OECD national accounts data files)

economy is contracting or expanding and if a threat such as a recession or inflation looms on the horizon. It is interesting to note that, despite our best efforts, mankind is not able to control the weather. Whether intended or not, this analogy raises the question as to the ability of governments and central banks to actually be able to control the economy, or if GDP is as intransigent to policy as the weather.

What Do We Know?

From the brief descriptions, we can conclude that the terms 'biodiversity' and 'GDP' have, in their respective disciplines, an overarching meaning from which policy decisions can be (and are) drawn at a number of levels. Both are used as a buzzword for the layman, high biodiversity and increasing GDP are seen as positive factors. Such statements remain unquestioned at a critical level, even by many professional commentators.

In the case of the term biodiversity, the accepted meaning is not supported by the hard data required to give it substance, it is a coverall for inadequate information enabling statements of apparent authority to be made in the absence of knowledge. In the case of GDP, the meaning has been stretched to cover implied factors which it doesn't measure, and it acts as a driver of policy without taking into account all of the implications of enacting such policies, it is too narrowly focused to enable it to do what it claims, so that authoritative statements have the potential for significant unintended consequences.

Both terms have one feature in common, they present an image of certainty whilst actually representing somewhat abstract concepts. These are sufficiently convincing to be considered as factual data, which mould attitudes and public perceptions of policy decisions.

These examples return us to the question of how much knowledge we actually have in respect of the truth of the environmental or economic sphere? There is a huge underestimate of the magnitude of the unknown in relation to both topics. Policies, plans and processes are regularly formulated which superficially appear to take into account the current ignorance of certain issues. However, when the magnitude of the knowledge gap is grossly underestimated, and the two topics coincide in a way which has major consequences, the capacity for error and misunderstanding is compounded, and the reconciliation process becomes a fudge of epic proportions.

It is often said that nature abhors a vacuum, that any available gap will be filled. I would assert that, in a similar fashion, the human mind also has a tendency to 'fill in the gaps' with regard to our knowledge and understanding of any observable phenomenon. It is a basic human need to be able to create an intellectual narrative which provides an explanation of the world around us, and the events which occur within it. Although such narratives may often be fanciful (myths and legends are prime examples of filling in the gaps in knowledge of a civilization) and are usually based upon a less than complete understanding of any given situation, the need for an explanation generally overrides the grey areas where uncertainty and ignorance predominate.

Alternative views of reality arise when there are spaces in knowledge and understanding to allow more than one interpretation of the existing evidence to be made. Some interpretations may be purely speculative, whilst others may be carefully constructed using detailed studies of the minutiae of the evidence available.

For example, many Victorian entomologists were great taxonomists and made observations which remain valid to this day. Many believed that these details were evidence of the omnipotence of a divine creator, whereas the data collected is equally as valid in support of the contrary view of Darwinian evolution. The facts and data collected are real and valid, but could be used to support either of two possible explanations for their existence as an observable phenomenon.

Similarly, economic data can be interpreted in many different ways, in fact there is a whole industry based upon making different interpretations of the same economic information, these viewpoints are sold to investors with the hope they will be able to use the latest interpretation of the data to take advantage through market speculation. More generally, economic data can represent the operation of a completely different system, whilst still broadly conforming to the view of the mainstream.

Mankind has a strong affinity with the identification of patterns, either spatially or through concepts such as cause and effect. Once a

Fig. 3   Duck or rabbit? (*Source* Jastrow [5])

pattern is identified it becomes very difficult to look at it from a differ-ent perspective. At a basic visual level, this can be seen in Fig. 3. Is the image a duck or a rabbit?

In more complex theoretically based studies, such as both ecology and economics, this perceptual bias (once described to me by a very skilled botanist as 'the eye of faith' when I was struggling to actually see what was described as glandular hairs on the stem of a rather sad and wilted plant specimen) remains the case, even if the perceived pattern of actions and results do not provide an optimal outcome and often fly in the face of further information which does not 'fit' with the per-ceived truth. Data tends to be either downplayed if contradictory to the received wisdom, or the prevailing model is grudgingly modified to increasing complexity in order to accommodate such supposedly anom-alous evidence.

An example of the development of ideas which adapt to increasing knowledge, leading ultimately to a complete paradigm change can be seen in the world view from early civilisations where the earth was the centre of the universe, to the later view of a heliocentric solar system within one of many galaxies.

Placing the earth at the centre of the universe is, in the absence of other data, completely understandable. From our perspective, we are on an unmoving and stable body and simple observations show that the other visible features in the sky appear to move relative to our fixed point of observation. Furthermore, the 'received wisdom' of scripture in many ancient cultures states factually that the earth is at the centre of the created universe. Early direct observations quickly revealed that, although the majority of the stars moved in unison around the earth, the sun and moon together with several other luminous bodies (planets) actually moved at different speeds. In order to account for the differences in observed motion, these bodies were placed upon a series of nesting crystalline spheres, each of which was rotating at a different speed, with an outer sphere holding the fixed stars. This model (the Aristotelian model) provided what appeared to be a rational and reasonable model of the universe, based upon the available data (Fig. 4).

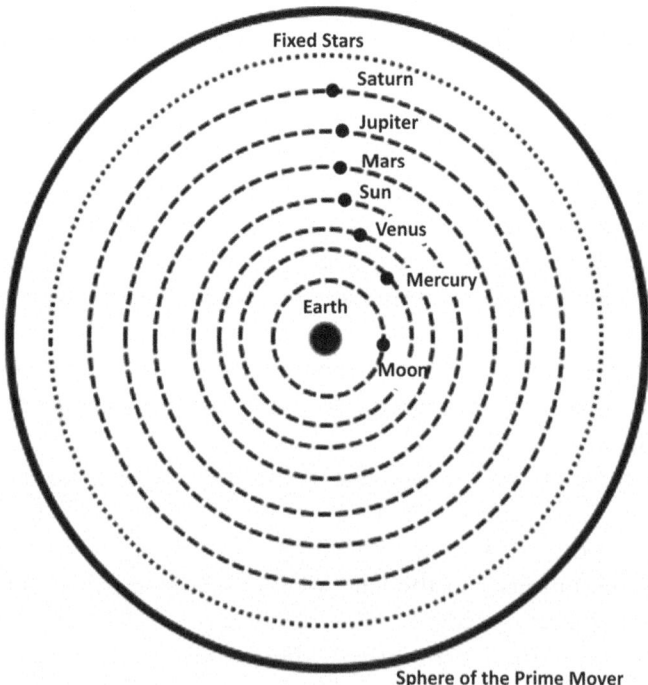

Fig. 4 Aristotle's solar system

Despite this general satisfaction, several anomalies remained. Particularly prominent in this were the recorded observations of planets apparently stopping in their orbits and then moving backwards (retrograde motion). These data required a rethink of the straightforward Aristotelian model. It resulted in a new system (the Ptolemaic model) which required the invention of hypothetical points of rotation around which each planet circled (epicyclic movement), as it continued its ongoing movement around the earth (which was now rotating around the physical centre of the universe rather than being the static point itself). This complexity was added to respond to the observed data, but was still not completely accurate. There appeared to be a stubborn refusal to allow for ideas which did not place the earth at the centre (Fig. 5).

It was not until the middle of the sixteenth century, following more than 1000 years of the Ptolemaic model dominating how the earth's place in the solar system was viewed, that a heliocentric model

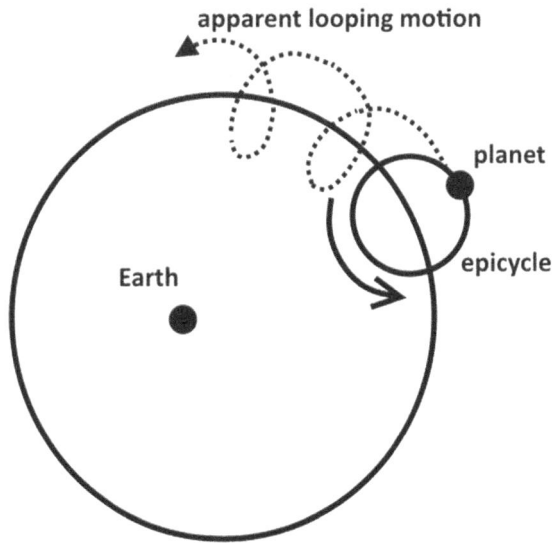

Fig. 5 Ptolemy's solar system

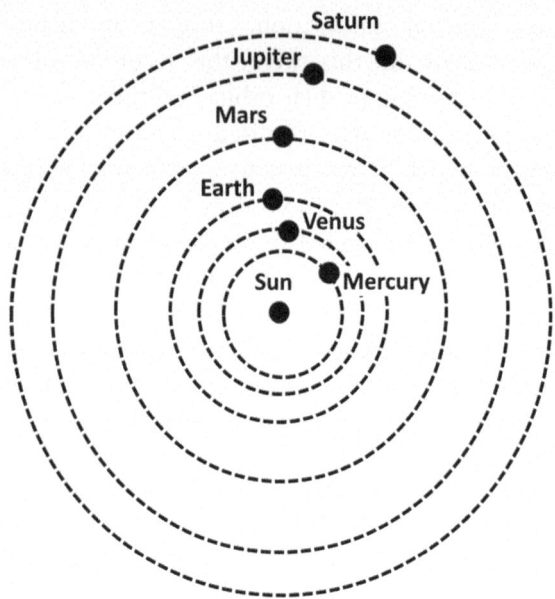

Fig. 6 Copernicus's solar system

was proposed by Copernicus. However, because it hypothesised circular orbits of bodies around the sun it did not offer a greater match to observed data than the Ptolemaic system. Both systems were widely held to be the correct model (Fig. 6).

The main 'logical' argument for a geocentric system, other than received wisdom and the word of scripture, was that the constellations did not alter shape or size as they rotated around the Earth. This was seen as the key fact for geocentric models, based upon an assumption that the stars are relatively close to the earth, and in a spherical shell. In reality, the assumption made is erroneous, the stars are much, much further away than could be imagined, and they do demonstrate relative movement, although this wasn't observable until instruments of sufficient accuracy were developed in the nineteenth century.

Once a heliocentric system had begun to establish itself in the group psyche, evidence began to be uncovered that supported this model. Much of this came from improvements in optical instruments which allowed more accurate observations to be made. Galileo recorded that Jupiter had moons which orbited it, thus undermining the idea that everything revolved around the earth. He also observed that Venus demonstrates phases, proving that it, at least, revolves around the Sun.

Johannes Kepler finally presented a system with the sun at the centre of the solar system and planets following elliptical (rather than circular) orbits which allowed him to successfully make predictions regarding observable phenomena. This finally established a heliocentric model as the 'mainstream'.

Through reflection on this example, the following insights can be drawn regarding how the human mind deals with an incomplete data set:

- Received wisdom can have a strong reinforcing effect upon a model. In this case, the statements of scripture were considered to be inviolable.
- The simplest model is that which is normally adopted first. Basic observation and instinct would place the earth as the unmoving centre, with other bodies rotating around it.
- Once adopted there is a tendency to increase the complexity of the original model to accommodate new information, rather than reject it and reconsider first principles. Seen in the development of the Ptolemaic system.
- Assumptions made at an early stage where there is a lack of data can become 'facts' in later iterations of a model, hence reinforcing its basis. The arguments regarding the lack of movement in the constellations, based upon the false assumption of their proximity to earth, illustrate this point.

By having such an inbuilt inertia and tendency to remain averse to truly new ideas, many significant advances in thought and technology have been delayed or suspended from widespread adoption due to this phenomenon. It is only when the accepted view of how the system works is in direct and overwhelming opposition to the evidence that a new paradigm is likely to be adopted.

The way in which the human mind deals with incomplete data, as described above, can be used to test how humanity has come to terms with unknowns and uncertainty in many different areas, including our knowledge of the environment and the mechanisms of economics. In this way, we now have some form of benchmark for a critical analysis of the current state of play. We can look at alternatives with a more open mind, as long as we are aware that the field is wide open and the 'mainstream view' is by no means in a position of certainty with regard to either the quantity or quality of, what appears to be, supporting information.

Of course, there is one major distinction to be made in comparing man's view of how the universe works with the natural world and the conceptual view of trade and economics. The way the universe is established has little effect upon our day to day lives, the data presents itself, but there is nothing that man can (yet) do to alter the seasons, the phases of the moon or our distance from other celestial bodies. When it comes to ecology, things are very different, man can, and does, frequently affect the way ecological systems operate, and if the 'model' he is adhering to is erroneous, then the consequences can be highly damaging, not only to the environment, but also to the aim of the intervention. For example, the ongoing drive to increase food production in Africa includes the promotion of agricultural mechanisation. However, there is potential for this to lead to a competitive advantage for those with larger landholdings, which would then acquire smaller, less efficient farms. This results in a loss of biodiversity, greater environmental stress on soils, water resources, etc., and an increase in rural unemployment, thereby increasing poverty.

When it comes to pure economics, management of the economy is seen by the ordinary citizen as 'the way it is'. There is a belief that the economy is a well-understood machine and the only differences are in how this should be operated. In fact 'the economy' is a result of predetermined interventions in financial and monetary systems based upon a model formulated as much by ideology as empirical data. In other words, the economy is viewed for some reason as more akin to a mechanistic, immutable and empirically well understood solar system rather than a poorly understood experimental test bed, which is closer to the reality of the situation.

Mistaken decisions or misdirected policy which have been brought into action as a result of errors in the models of the environment and economic theory used to formulate decisions can have a large impact on everyday life, with significant risks of damaging both the environment and quality of life. The debate surrounding the place of the earth in the solar system pales into insignificance when compared to the potential real-world consequences of a mistaken environmental or economic world view. Due to this, it is even more critical that we have the fullest available knowledge of these subjects. Once it is realised that decisions are based on a best guess approach, it is certainly of benefit to keep an open mind about whether or not existing received wisdom is, in fact, true.

Highly detailed studies, which lead to the development of complex models, may, superficially, appear to offer the best way of providing the most effective decision-making tools. However, there is a danger here in making the assumption that a model derived from a highly detailed study is the most generally applicable, as it may only be truly applicable in one, highly specific set of circumstances. If it is then extrapolated to apply to one or more alternative paradigms, the basis of the study does not necessarily support its new application. Such transference of studies are frequently described as modelling or innovative application into new fields, but in many cases, it is merely the cut and paste use of detailed situational research into an arena where its effect is unstudied and untested.

There is a general view that it is better to do something, even if based upon partial information or data with only passing relevance to the situation at hand, rather than do nothing. Although this represents a reasonably good working definition of risk taking, it is worth bearing in mind that under certain circumstances uninformed action may actually serve to make the situation considerably more difficult, especially if it is framed in a plausible narrative.

These errors are somewhat analogous to the procedure adopted in the solving of a Sudoku puzzle. Initially, there are many possible numbers which can be inserted into any given box without causing the puzzle to fail. Finding the actual correct number is a more time consuming and (some would say) tedious process of extrapolating each possibility until only one number can be correctly placed in any given box. Without adopting a careful approach all seems well initially, number can be placed in locations which appear to be correct (or at least not breaking the puzzles rules) but eventually errors using this 'scattergun' approach reach a point where suddenly the whole process seizes up and no number will fit without breaking the format of the puzzle. At that point to try and retrace steps is a futile process, and it is far better to start again, with a much more careful approach.

Although this can provide a useful lesson regarding the need to be methodical for the individual completing the puzzle, the idea of the 'scattergun' approach to modelling either economics or ecosystems (using the maxim 'if it hasn't failed yet then our decisions to date must all be correct') runs the risk of delivering either a total economic or ecological reset, neither of which are as inconsequential as turning the page in a book of puzzles.

The information used in the development of a model has two fundamental characteristics, quality and quantity. Both are significant elements when trying to construct a model which is applicable to a range of situations.

When considering the quality of information it is important to consider both the circumstances surrounding the collection of data (in terms of, for example, sample population demographics or habitat variability) and also the likely end use of a model. For example, a model of agricultural patterns and economics based upon a lowland system in western Europe will become increasingly inaccurate as the type of agricultural practices and prevailing economic system varies in other regions throughout the world. Any failings in the quality of data collection can even cause a model to be highly inaccurate within the precise conditions it was formulated within.

Quantity of data also has a great deal of influence on how a given set of circumstances is interpreted. With small amounts of data, a greater number of possible underlying mechanisms to explain the observed results can be postulated. As the amount of information increases, it becomes increasingly possible to identify the general principles underlying observed phenomena. There is, however, an important caveat to this as an ongoing process. Any existing hypothesis must be freely abandoned if data begins to show the model is not meeting its required function. Unfortunately, there is a considerable amount of inertia involved in this process, and data may be dismissed as an outlier, the methodology of new data collection may be questioned (see the data quality issue described above) or the aberrant data may be put down to a system which is failing to function 'correctly' (whereas it is our understanding of the function which is incorrect).

Whatever the reason things fail to behave as they are 'supposed to', humanity always manages to provide a plausible explanation. Changing an entire world view is usually considered as a last resort. This makes perfect sense in a world where reputations and careers can be totally undermined if the assumptions upon which expertise is based become defunct. As a result of this, the unknowns in both the fields of economics and ecological science are instinctively filled with a narrative to explain why this lack of knowledge in certain areas is of no significance in maintaining the existing paradigm. This is a perfectly natural response, but if society makes assumptions about the unknown, underestimating the levels of uncertainty, and proceeding with a confidence which is based upon a partial or distorted view of the situation in hand there is always the potential to be just plain wrong in the decision-making process—and this can be a major problem.

About the Book

This book does not promote any particular political narrative regarding the state of the economy and the environment. It aims to offer an alternative understanding of how our 'known unknowns' in these fields may be interpreted. It presents a plan of how, by accepting that the commonly held view of way the world works isn't necessarily the only way, it is possible to develop a system which avoids the constant clashing of heads between human commercial activity and the environment.

At its basis, this book presents an exploration of the following hypotheses:

- That the concept of Ecosystem Services as presently formulated is embedded within an economic system which does not offer the best outcomes for individuals, society or the environment.
- That current Ecosystem Services thought and theory is complicit in the maintenance of an erroneous economic system which serves to merely obscure a significant failure of current economic theory.

An alternative approach to environmental valuation is proposed which offers a more creative and constructive viewpoint. Its adoption could serve to lead to a more practical and rational economic system which is more respectful of the needs of all stakeholders (sentient or otherwise).

The remainder of this book is structured in 5 chapters, each of which aims to provide an individual narrative on a particular aspect of the subject. Inevitably, however, these are intertwined to present a view of how Ecosystem Services, ecology and economics are related and how the development of a new way of looking at our relationship with the environment and our day to day activities can influence policies which affect us all.

Chapter 2 "Basics" is concerned with providing an outline of the current state of play with regard to the underlying theories and historical development of the Ecosystem Services concept. It then presents an outline of the current economic system, showing how this has been used as a foundation for the Ecosystem Services concept as currently formulated. It then raises questions regarding how well there is a fit between economics and the aspirations behind the development of Ecosystem Services through the introduction of alternative economic views. The chapter concludes by looking at how well Ecosystem Services as an idea can be validated by alternative economic theories.

In chapter 3 "Parallels and Function" a description of how similar terms employed by ecologists and economists may have entirely different functional meanings is presented. The aim of the chapter is to identify potential sources of misunderstanding when specialists from each discipline begin to communicate using the concepts inherent in Ecosystem Services.

The range of methods used to value ecosystems is reviewed in chapter 4 "Valuing Ecosystems". It then proceeds to critically assess how value is defined in mainstream economics and offers alternative viewpoints of the concept of value based upon other economic models.

Chapter 5 "A New Model" draws together the issues identified in previous chapters and sets out to establish an alternative model for bringing together the environment and economics paying particular attention to both the concept of Ecosystem Services and the potential to embed environmental issues firmly within the financial framework.

Chapter 6 "Effects and Applications" explores how the application of the alternative model would affect the implementation of Ecosystem Services principles upon stakeholder groups, including ES practitioners, planners, business and financial interests and citizens.

References

1. http://www.catalogueoflife.org/col/info/ac. June 30, 2017.
2. Mora, C., Tittensor, D. P., Adl, S., Simpson, A. G. B., & Worm, B. (2011). How Many Species Are There on Earth and in the Ocean? *PLoS Biology, 9*(8), e1001127. https://doi.org/10.1371/journal.pbio.1001127.
3. Rowson, B., Anderson, R., Turner, J. A., & Symondson, W. O. C. (2014). The Slugs of Britain and Ireland: Undetected and Undescribed Species Increase a Well-Studied, Economically Important Fauna by More Than 20%. *PLoS ONE, 9*(4), e91907. https://doi.org/10.1371/journal. pone.0091907.
4. Samuelson, P. A., & Nordhaus, W. D. (2012). *Economics* (19th ed.). New Delhi: McGraw Hill Education.
5. Jastrow, J. (1900). *Fact and Fable in Psychology*. Houghton: Mifflin.

Basics

Definition and Historical Development of Ecosystem Services

Human Dependence on the Environment

Modern life has, in many ways, detached human activity from the natural environment around it. This is particularly acute in the urban environment (where the vast majority of people live) resources from outside are generally supplied to the urban dwellers and there is little interaction with the natural world for the basics of survival (food, shelter).

However, developments in information sharing and media coverage has highlighted to the urban dwellers in the more economically prosperous population centres the environmental degradation, habitat loss and pollution occurring in the wider environment.

This has presented a dichotomy, where the rural poor in fundamentally agrarian areas, with less access to media coverage and information sharing technologies are less aware of the wider status of the environment. Their main concerns are those changes which directly impact

© The Author(s) 2019
S. Muddiman, *Ecosystem Services*,
Palgrave Studies in Natural Resource Management,
https://doi.org/10.1007/978-3-030-13819-6_2

upon their livelihoods and means of survival, that part of 'the environment' which they interact with as part of the fabric of their lives. In contrast, the more detached populations are far more conscious of global and widespread change.

This awareness of the global environment has spawned many initiatives and approaches to environmental protection in a top-down approach, manifested in a range of forms such as regulation, pollution control, and the legal protection of habitats and species. Although much of this control has had the desired effect, it has had a tendency to overlook the sensitivities of those who are closest to the issues being debated.

Industrial activities have in the past been curtailed in the name of environmental protection without any reference to the local loss of employment it engenders. Landowners have had their livelihoods controlled through the designation of their land as protected without their consent or consultation. For example, in June 2017 the California Farm Bureau and two ranchers' associations sued the U.S. Fish and Wildlife Service, challenging a decision to designate more than 1.8 million acres of rural California as 'critical habitat' for three species of amphibians protected by the Endangered Species Act. The lawsuit asserts that the critical habitat designation subjected farmers 'to substantial regulatory burdens' which could put ranchers' livelihoods at risk [1].

It is a result of this ideological divide, between environmental protection in all of its forms and the efforts of people to conduct economically advantageous activities, which has brought about the need for a framework which not only allows for the protection and preservation of the environment, but also allows the freedom for dependent communities to conduct economic activities which alter it.

It is without doubt that industrialisation leads to significant environmental change. Much of this is piecemeal in nature, as a particular technology develops it spawns a range of 'projects' which continue to alter the status quo. For example, the development of the internal combustion engine and the automobile not only led to a surge in petroleum sourcing, car factories, steel production, etc., but also generated government-sponsored infrastructure to allow the newly created transport to operate across continents. This industrial growth was not planned as a single entity, rather it developed in response to economically driven consumer demand and supportive government policy.

This type of technologically driven change resulted in a mindset that things were 'done to' the environment by mankind, and that it held dominion. Nature was being tamed by the achievements and endeavours of an industrialised society. Of course, society was not immune from the effects of their actions. In London, a high population density together with the widespread use of diesel-powered vehicles, coal-fired power stations and the burning of low-grade coal, combined with particular climatic conditions occasionally led to the production of a dense acidic fog (smog). Most notable of these events occurred in 1952 where a dense smog occurred which is thought to be responsible for up to 12,000 mortalities [2]. This event was directly responsible for the introduction of clean air legislation in the UK.

Despite these issues, for the urban industrialised societies which these advances had created, humanity was on an upward trajectory. Any incidental problems would likely be solved by appropriate technological development in the very near future.

In opposition to the idea of mankind's dominion over nature, a more holistic view of the relationship between man and nature was developed. This is most readily expressed in the works of James Lovelock and Lynn Margulis, through the Gaia hypothesis [3]. This proposes that the living component of the earth (the 'biosphere') has a synergistic and self-regulating relationship with the non-living components of the environment, to form a complex system which maintains conditions suitable for life.

Apart from some philosophical criticisms of the theory, such as the implication that the planet is in some way 'directing' its regulatory feedback, there is one significant flaw in the concept. In considering the homeostatic and balanced view of the world presented in this hypothesis, the activities of mankind are at best ignored and at the most extreme considered to be an active interference in an otherwise functioning system. This is, therefore, merely 'the other side of the coin' in that it, like the mindset of industrialisation, considers mankind to be in some way distanced from the environment in which it operates.

This is clearly a significant contradiction when attempting to construct a holistic view of the biosphere (which must, by definition include mankind and its activities). By making human actions an exception, this approach either denies the place of humanity within the global system, or is a self-fulfilling rebuttal of the whole concept. Because, if mankind is an integral part of the whole biosphere, then how can its influence be anything other than a part of the regulatory mechanisms of the whole?

One further corollary of the idea that humanity is in some way a separate (and generally deleterious) influence on the planet, is that it opens the door for interventionism on a grand scale. The reasoning being that if human actions are the cause of the 'problems' then it is necessary to act again, in opposition to these damaging actions, to correct the system as a whole. It places mankind and its actions, guided by whatever motivation, as the fundamental governing force in the well-being of the environment.

The contrasting viewpoints of the relationship between mankind and the environment both share the view that mankind is separate from the rest of the natural world in terms of its activities. The commercialised 'dominion' oriented view is that the Earth represents a series of resources which are available for use in order to further human progress. The 'Gaian' view in reality sees mankind as an ignorant or errant product of a holistic system, which although it acts as though it is outside of the natural complex of interactions its activities can be seen as little more than acts of self-harm. These polarised worldviews bring into sharp relief the need for a balance to be reached which acknowledges the interrelationship and integration between society and its surroundings. There is a clear need to reach a compromise situation, one which acknowledges our dependence on the environment, but which also accepts that there are areas of this relationship where sourcing and use of natural resources benefits us as a species. The concept of Ecosystem Services is an attempt to achieve such a balanced view.

Millennium Ecosystem Assessment

Much of the development of Ecosystem Services, as currently framed, has been derived from the framework initially presented in the Millennium Ecosystem Assessment (MEA) [4]. This project involved the work of more than 1360 specialists and assessed current knowledge, scientific literature, and data, comprising a synthesis of existing information at the time of compilation (2001–2005) rather than being new research. The study was instigated to look at ecosystem change and to consider the consequences of such change on human well-being.

When defining ecosystems, the MEA describes a number of very broad categories such as 'woodlands', 'coastal' and 'drylands'. The majority of the terrestrial ecosystems in the MEA derive from the prevailing physical conditions and represent a coarse-grained version of the Holdridge life zone classification [5]. This system allows soil type and climax vegetation to be predicted by the prevailing climate, altitude and latitude. In the MEA, the remainder of the terrestrial land mass is classified as either urban or cultivated land, i.e. anthropogenic in nature. Rather than being functional ecological units, these 'ecosystems' can be more accurately defined as a series of biophysical types which act as receptor/provider systems for human activity.

One of the key propositions of the MEA was that people are an integral part of ecosystems, so a dynamic interaction exists between them and other parts of the environment. This approach echoes the view of the holistic 'Gaian' approach, but goes one step further, by embedding human activity into ecosystems as part of its intrinsic characteristics.

Although humans are a widespread species and can certainly be considered to have at least an indirect influence over all parts of the globe, the assertion that humans are an 'integral part' of an ecosystem is only true in those rare cases where a human population lives in isolation and has achieved a sustainable state with its environment without any ongoing directional anthropogenic change. It is only under these circumstances that the definition of an ecosystem (as used in the MEA): 'A dynamic complex of plant, animal and microorganism communities and their non-living environment interacting as a functional unit' remains valid.

By including humans as a whole in this 'functional unit' it is necessary to stretch the definition of 'ecosystem' to cover the entirety of life as a single entity, as human activity represents a complex of global interactions which extends across all ecologically defined biotic boundaries.

This view runs counter to the observed evidence that, by and large, human activity leads to changes in ecosystems which are not in any form of equilibrium. This is equally true for the damage caused by resource exploitation as it is for conservation management activities which require ongoing, planned interventions to achieve an appearance of stability. This perspective, of man as some form of external agent, therefore applies equally to both the 'Dominion' and the 'Gaian' views of the human relationship to the earth.

Whether the effect is either moral or rational is the point of divergence, rather than whether or not mankind is acting upon the environment in a 'natural' manner. Human consciousness inevitably serves as a barrier between the inner self (that which perceives) and the outside world, including the environment (the subject of perception). Even if activities are immensely destructive, they cannot be considered to be unconscious. They can, however, be considered uninformed (or misinformed).

In addition to considering our relationship with ecosystems, the assessment was charged with identifying actions for the enhancement, conservation and sustainable use of ecosystems. This element of the project clearly, and in a somewhat contradictory manner, presents the role of mankind as an active, external influence upon the natural environment rather than being considered an integral part of it.

In terms of valuing the environment and ecosystems, the MEA asserts that the actions people take which maintain ecosystems result from the 'intrinsic value' of species and ecosystems. This intrinsic value is defined as the 'value of something in and for itself, irrespective of its utility for someone else'. This concept is something which is true only in relation to the motivation of the direct action involved. As a concept with broader applicability, it is purely philosophical in nature as it cannot be quantified or necessarily predicted. It is not an economic valuation in an analytical sense, although the statement sounds superficially to be a description of an economic reality.

Outside of actions of individuals, any action undertaken requires the expenditure of time, labour and materials, all of which fall firmly in the economic sphere and reflect directly the value which society places upon the tasks to be undertaken. Even if payment is not made directly (such as works undertaken by charitable organisations) the accumulated wealth that generated the ability of society to fund the work also has an economically quantifiable value. Even when tasks are undertaken on a purely voluntary basis, travel, accommodation and subsistence are valued and paid for. In addition, motivations for voluntary activity in the environmental sector are rarely purely altruistic, opportunities for travel, enhancing future employment, social status amongst peer groups and the 'feel good factor' are major motivations for many volunteers.

When considering the concept of human well-being, any measures will be subjectively selected and not easily quantified. In the MEA, human well-being does not have a single definition, but rather is considered to comprise five main elements:

- Security, which is divided into personal safety, secure resource access and security from disasters.
- Basic materials for good life, comprising adequate livelihoods, sufficient nutritious food, shelter and access to goods.
- Health, comprising strength, feeling well and access to clean air and water.
- Good social relations, made up of social cohesion, mutual respect and the ability to help others.
- Freedom of choice and action, which is defined as the opportunity to be able to achieve what an individual values doing and being.

The MEA avoids relating human well-being to economic growth and development in this compound definition. This acknowledges that GDP as a metric does not have a direct relationship to well-being, as mentioned in chapter 1 "Introduction".

The concept of Ecosystem Services is introduced in the MEA as a framework to link ecosystems and human well-being. They are defined in the MEA as the benefits people obtain from ecosystems, which include *provisioning services* such as food, water, timber and fibre;

regulating services that affect climate, floods, disease, wastes and water quality; *cultural services* that provide recreational, aesthetic, and spiritual benefits; and *supporting services* such as soil formation, photosynthesis, and nutrient cycling. The MEA identifies humanities fundamental dependence upon the flow of these Ecosystem Services.

It should be borne in mind that the benefits of ecosystems to humans, via Ecosystem Services, tend to be only realised when ecosystems remain unaffected by human action. In other words, human action is not an integral part of the ecosystem function, rather humanity is very much in the role of consumer of services provision which has come about through a serendipitous interaction of the non-human elements of the environment. Due to our lack of information and knowledge with regard to ecosystem composition and function, humanity, at best, can be considered a sustainable exploiter of ecosystem function, and at worst a destabilising influence on a system of interactions which we do not understand. In our current state of knowledge, our management of the environment can best be considered the proverbial 'bull in the ecological china shop'.

In conclusion, the MEA identified a recent rapid change in ecosystems, which it assigns to human activity, primarily in terms of increasing demand for food, fresh water, timber, fibre and fuel (through activities such as industrialisation of agriculture, international trade and energy sourcing). Furthermore, it asserts that these changes have resulted in a substantial and largely irreversible loss in the diversity of life on Earth. However, as we have seen in chapter 1 "Introduction", there is only a very limited understanding of the levels of diversity which exist, so the foundation for an assertion of this type can only be considered to be speculative at best.

The report acknowledges that changes in ecosystems have contributed to substantial net gains in human well-being. However, it goes on to acknowledge that there are also costs through the degradation of many Ecosystem Services, the increased probability of further changes as a result of knock-on effects, and increased poverty for some groups of people. This implies that human well-being is considered in a global, aggregate sense, rather than in terms of individuals or communities. Unfortunately, this conclusion leads to a vision of ongoing ecosystem change as long as the bulk of humanity benefits from it, whilst sacrificing the well-being of those who are perhaps most closely associated with the systems being affected, but without the authority to prevent damage to their own communities and livelihoods.

The MEA recognises that reversing the degradation of ecosystems whilst meeting increasing demands for services will involve significant changes in policies, institutions and practices that are not currently under way. This recognition maintains the concept of a 'top-down' approach to our relationship with the natural world. Rather than having humanity as an integral part of ecosystems as originally asserted, the MEA has shown that humanity in general does not seem to be capable of being integrated with ecosystems. It needs a hierarchical guidance mechanism which engages from a detached and theoretical standpoint whilst directing the activities of those who are most intimately connected with the Ecosystem Services which are being affected.

One factor omitted from the brief of the MEA was the evaluation of the economic and policy drivers which lead to ecosystem changes. This is a major gap, as many of the actions of mankind which cause observed ecosystem changes are a response to the prevailing economic system. For example, the generally accepted view that any nation should strive towards growth as defined by an increasing GDP, or that nations' borrowing for development purposes is financially prudent.

TEEB

The Economics of Ecosystems and Biodiversity (TEEB) is a global initiative focused on 'making nature's values visible'. Its principal objective is to 'mainstream the values of biodiversity and Ecosystem Services into decision-making at all levels' [6]. Its primary aim is to develop a valuation system for biodiversity and Ecosystem Services which can be used to inform policy. The initiative began in 2007, and it adopted the framework of Ecosystem Services devised for the MEA.

At its inception, the process was driven by the environment ministers from the G8+5 countries. It was proposed in order to analyse the global economic benefit of biological diversity, the costs of the loss of biodiversity and the failure to take protective measures versus the costs of effective conservation.

Of particular note is the inclusion of the term 'biodiversity' in the initiative. As we have already seen in chapter 1 "Introduction", this is a very poorly understood area, and without a significant improvement in our knowledge, it is almost impossible to draw meaningful conclusions. Based upon the fact that we only have knowledge of 10% or less of the global biodiversity resource, it becomes clear that this project is fundamentally an effort driven by policy leaders. The aim is to develop some form of economic quantification of ecosystems and biodiversity without fully engaging with the knowledge deficit in terms of species and their interactions. This is, by necessity, a broad-brush approach tailored for decision makers and based upon the concepts of a hierarchical top-down decision-making process working within the current economic framework, which is detailed later in this chapter.

The first stage of the project resulted in the production of an Interim Report [7] which presented a narrative for the basis of the project and set the context for further phases of work. Following this a second stage of the project was initiated, which resulted in the production of four key publications, together with a Synthesis Report [8] which presented a synthesis of the following studies:

- TEEB Ecological and Economic Foundations: A report on the fundamental concepts and state-of-the-art methodologies for economic valuation of biodiversity and Ecosystem Services [9];
- TEEB in National and International Policy Making: A report providing analysis and guidance on how to value and internalise biodiversity and ecosystem values in policy decisions [10];
- TEEB in Local and Regional Policy: A report providing analysis and guidance for mainstreaming biodiversity and ecosystem values at regional and local levels [11]; and
- TEEB in Business and Enterprise: A report providing analysis and guidance on how business and enterprise can identify and manage their biodiversity and ecosystem risks and opportunities [12].

Following the production of these reports, the project proceeded to apply the findings of the studies in two areas. The first, termed 'Country Projects' seek to build national, regional and local government capacity in the production of economic assessments relating to the valuation of ecosystems and biodiversity, and also to translate this information into policy-making activity. Secondly, analysis of specific sectors and biomes is being undertaken with the intention of identifying the sources of value in biodiversity and ecosystems.

Although offering what appears to be a logical way forward in the overarching aim of integrating economics and biodiversity, these key areas contain several conceptual pitfalls worthy of consideration at this stage.

Of necessity, the production of economic assessments is embedded within the current economic system which, as we shall see, is a far from accurate model. Adherence to mainstream economics leads to distortions of the values assigned to ecosystems and biodiversity. This is fundamentally because of a top-down methodology which often fails to take into account the relative value which a resource has for the individuals who are most intimately associated with the Ecosystem Service itself (and who are often fiscally disadvantaged and without any control or power over the use or ownership of the land involved). In addition, the valuation is undertaken using a purely fiat monetary system which is debt based and largely abstract in nature. Using this approach identified costs can be met through increases in the money supply at the level of Central Banks, or the taxation of citizens and/or businesses. In the use of such a balance sheet, the loss of environment and dependent livelihoods or the disruption of communities can be demonstrated to be an economic 'price worth paying'. Because of this economic system, there is no way that an Ecosystem Service can ever reach the level of being considered too valuable to be destroyed based upon price alone.

As a result of an economic valuation in which any cost (in terms of monetary value) can be met, there is a need for additional policy to be formulated which confers legal protection to habitats and species. This actually bypasses the purely economic valuation process. However, the lack of knowledge regarding biodiversity, as outlined in chapter 1

"Introduction", presents a risk that legislative protection is not accurately formulated or focused.

One of the characteristics of ecosystems and biodiversity (as far as it is understood) is that they do not respect national borders. A valuation based upon a geopolitical division will inevitably overvalue features which are not well represented in that area, even if they are abundant elsewhere and vice versa. This again can lead to distortions not only in the valuation procedure, but also in the selection of features and degree of legislative protection provided by a national government.

The TEEB approach, of considering Ecosystem Services offers a means of overcoming some of these pitfalls, as it is functional utility which is being valued rather than any level of perceived rarity (when it can be determined with any level of accuracy). However, the lack of a clear distinction between Ecosystem Services and the underlying biodiversity which generates them presents its own series of difficulties. Although TEEB focuses upon the services aspect of ecosystems, it fails to fully acknowledge that it is virtually impossible to extract the 'service' element from the underlying biodiversity without a full knowledge of how the services are delivered. Such analysis requires a much more detailed understanding of the composition of the ecosystem than we have available at the present time. It is here that the concept of Ecosystem Services collides with the biodiversity knowledge gap.

Without understanding the mechanisms of delivery, the service being assessed is only crudely appreciated. This leaves a significant risk of misdirecting effort to those areas of services which appear to be the most obvious. Without the requisite detailed understanding of ecosystems and their individual components placing the 'cart before the horse', is a constant threat. By failing to fully understand the ecosystems in question and treating them as black boxes, with only a superficial knowledge of the operations going on inside, whilst at the same time tinkering with the lesser understood components through a range of permitted activities there is a high likelihood of unintended consequences.

The Four Categories of Ecosystem Services

In both the MEA and TEEB a commonly adopted classification of Ecosystem Services is used as a vehicle to enumerate and evaluate the value of the natural environment to man (Fig. 1).

Provisioning Services

Provisioning Services describe material or energy outputs from the natural environment. This category includes any material sourced directly from the environment including food, water and raw materials. Most Provisioning Service resources are relatively easily valued as commodities and are generally already visible within the commercial sphere through well-developed markets.

When considering the provision of food, TEEB recognises that managed agricultural systems provide the bulk of such resources. It identifies several sources of non-agricultural food as provisioning services, namely marine, freshwater and forest systems. In such 'natural' systems there is generally some form of human intervention, to a greater or lesser

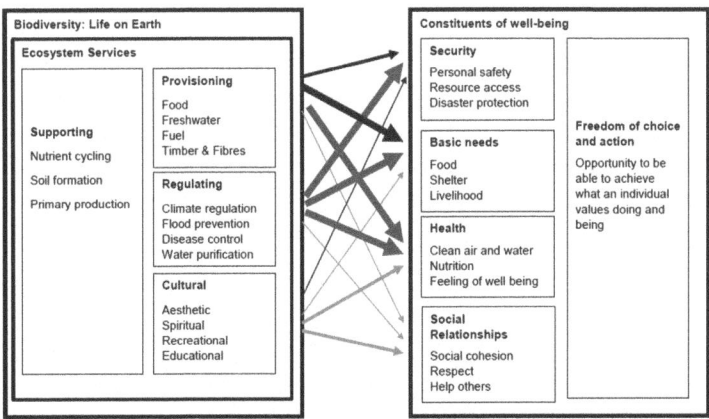

Fig. 1 Classification of Ecosystem Services (*Source* Millenium Ecosystem Assessment). *Note* Darkest arrows show greatest potential for mediation by socio-economic factors. Widest arrows show greatest intensity of linkage

degree. This may include habitat modifications such as clearing vegetation around open spaces where potential food resources gather restricting the times of the year when a particular prey item may be hunted, or the selective removal of competing species. Such management activities prioritise maximising the sustainable yield of the valued products over the maintenance of the system as a whole. Historically, such selective management of wild resources is likely to have been the precursor to full-blown agricultural activity. Firstly, a resource would have been recognised, then it would have been managed in situ, finally it would have been removed from the extraneous ecosystem and maintained as a crop or herd in its own right. The species most amenable to such domestication have provided the key species used as food sources for the majority of historical time.

The second Provisioning Service identified by TEEB are raw materials such as wood, biofuels and plant oils. Again, it should be emphasised that selective management and cultivation of such resources is generally adopted once a valuable resource has been identified. Very few such materials are sustainably sourced from a natural ecosystem over an extended period of time. Although it is true to say that these products were originally part of an ecosystem (which is true of all materials according to the definition of ecosystem used) the maintenance of the ecosystem is generally divorced from the production of such 'legacy' materials to a large degree.

When considering accessible (above ground) fresh water as a provisioning service, there is something of a logical 'chicken and egg' paradox presented. Ecosystems are described as playing a vital role in the global hydrological cycle. It is, of course, also true to say that the global hydrological cycle plays a vital role in the location and types of ecosystems which occur. As such, fresh water could equally be described as a cause of ecosystems as much as a service provided by them.

Ecosystems and biodiversity provide many plants used as traditional medicines as well as providing the raw materials for the pharmaceutical industry. The inclusion of medicinal resources as a provisioning service emphasises the lack of knowledge which we currently have regarding the pharmaceutical potential of ecosystems, as only 10% of all species are identified. Chemicals from this small proportion of biodiversity are responsible for the majority of medicinal products, either as a direct extract, a chemical starting point for further modification or as a chemical template for subsequent synthesis. As our knowledge is so limited in this area we must conclude that all ecosystems contain species which are a potential source of future medicinal resources.

One further provisioning service not specifically addressed in TEEB is the politically difficult issue sourcing of animal and plant products considered precious for cultural and traditional reasons. Examples include ivory, rhinoceros horn and scarce tropical timbers. In these cases, high rarity value leads to higher prices for the goods. In turn, this leads to increased risk-taking and an unsustainable use of the resource involved. Because the resource usually becomes an 'heirloom item' rather than being consumed as a commodity there is an economic tendency to maintain a limited supply rather than engage in any form of sustainable management. These resources are treated as an extractive industry. In fact, mainstream economics would consider the extinction of elephants and rhinoceros to be of economic benefit to those current owners of such resources, as their value would increase significantly if supply was permanently curtailed. Again, legislative measures are required to curb the tendencies of mainstream economic theory.

With the exception of fresh water, all of the above Ecosystem Services can be considered to be organism based, in that they are excess 'production' or by-products of individual species or groups of species occurring within any given ecosystem which can be extracted, cultivated and made use of by man. They are susceptible to human management and isolation for cultivation and can often be removed from their ecosystem context. Therefore, the maintenance of an ecosystem to support them is not usually critical to their production in the long term.

Regulating Services

Regulating Services are the services that ecosystems provide by moderating fluctuations in the environment. They can be considered to act as a negative feedback mechanism, serving to dampen extremes and providing more stable environmental conditions, e.g. regulating the quality of air and soil or by providing flood and disease control.

In most cases, the magnitude and value of regulating services are only recognised once they become non-functioning. This is a very different concept from the provisioning services described above, where there is a definitive product to be valued. To undertake an economic valuation of regulating services it is necessary to take an alternative approach, that of replacement costs. For example, if an outbreak of disease occurs as the result of a breakdown in regulating services, it would be necessary to consider the cost of enhanced medical provision and pharmaceutical intervention (assuming such intervention is available). An increase in flooding would be valued via the cost of engineered flood prevention and protection measures.

Regulating services can conveniently be divided into two distinct functional services (from the perspective of humanity as the consumer of such services).

The first class of regulating services are those which act to stabilise or moderate the physical conditions of surrounding areas through the operation of a fully functioning ecosystem. In these cases, there is no active benefit to any specific human activity dependent upon the continuation of the service. Such services act as a background mechanism upon which a diverse range of activities take place. Such services include water purification, flood defence and air quality moderation. In these cases, the natural ecosystem and the modified systems resulting from human activity are separate, with the latter benefitting from the former without any conscious intervention. In addition, there is no individual biological element which provides the service. It is dependent on the holistic function of the ecosystem. To coin a phrase, such mechanisms could be termed intersystem effects, i.e. the service is provided from one ecosystem to other, unrelated receptors.

The second form of regulating services occur where man has modified an ecosystem through conscious manipulation and management (such as the selection and cultivation of native plants), but that 'stripped down' system is still reliant upon elements of the unaffected, original, ecosystem in order to function. This incidental application of internal ecosystem activities to areas where a non-integrated, but partially related system, has been modified by previous human activity can be termed and intrasystem effect. For example, the pollination of agricultural crops dependent upon the presence of original pollinators, although they occur only in undisturbed ecosystems or agents of biological control which extend from undisturbed habitats when their prey or host species become abundant in adjacent agricultural systems.

It is worth noting that certain biological control agents have been isolated, bred and commercially traded. Therefore, biological control agents could equally be considered to be a provisioning service (see above). This resource offers possibly as great or greater potential than medicinal provisioning services, once knowledge of species and their interaction has been achieved.

The two forms of regulating service are shown in Fig. 2.

This distinction is of significance particularly in relation to valuation methodologies. Intersystem services are generally replaceable through the provision of alternatives which are not part of the regulating eco-system, e.g. engineered solutions, pharmaceuticals, alternative habitats, etc. Intrasystem service replaceability is much more concerned with the isolation of key species in existing ecosystems and drawing them into the sphere of semi-domesticated culture or cultivation.

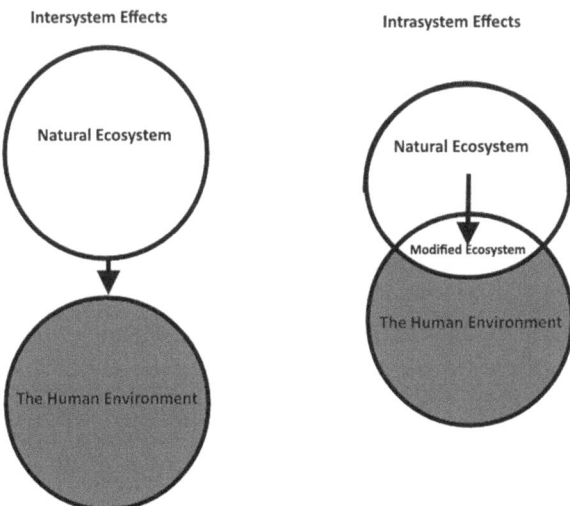

Fig. 2 Types of regulating service

Habitat or Supporting Services

These services represent the processes which occur within an ecological system which serve to maintain its sustainability. Such 'services' include soil formation, photosynthesis and nutrient cycling.

The introduction of the concept of habitat or supporting services within the TEEB classification presents something of a circular argument. In ecological terms, habitats and the processes they perform are themselves part of ecosystems and are not separable from the ecosystem as a whole. In the same way, individual species are integral parts of habitats, and it cannot be said that habitats 'support' species any more than ecosystems 'support' habitats.

It is somewhat similar to concluding that the organs of the human body are separate and the purpose of the body is to support its constituent organs and their functions. For example, the human lung absorbs oxygen and emits carbon dioxide, this function is necessary to support the body which supports the lungs which provide this very action. But it is a little unreasonable to suggest that the purpose of the human body is to absorb oxygen and emit carbon dioxide.

When it comes to the maintenance of genetic diversity in its function as an Ecosystem Service, TEEB presents two elements which are quite distinct in ecological terms.

Firstly, there is the phenomenon of genetic drift and adaptation of species to their environment over time. This is considered to be a potential (but unknown) benefit through providing resources which could lead to the development of cultivars and new breeds of domesticated sources of food. In this respect, the gene pool present in the non-exploited environment represents a potential Provisioning Service.

The second element is a reflection of the distribution of biodiversity around the globe. Some habitats have an exceptionally high number of species which makes them more genetically diverse than others ('biodiversity hotspots').

The implication of this is that as most diversity is assumed to be concentrated in a few areas (as far as our limited knowledge of biodiversity is concerned) these areas represent the most important areas for potential future discovery and use of genetic resources of value to humanity. Unfortunately, this does not necessarily represent the truth of the situation.

One consideration which has not been taken into account in this service is the potential utility of the genetic resources present. With our current state of knowledge, there is no reason to believe that a habitat with few species does not contain genetic resources which may have great potential utility to mankind. Nor is there evidence to suggest that single genetic resources extracted from a highly diverse ecosystem are inherently more valuable.

As there is no direct utility (from an anthropocentric point of view) in relation to habitat services (i.e. value ascribed to their existence rather than their function) the arguments relating to their valuation relate to the concept of intrinsic worth (and the presence of future exploitable genetic resources, currently unknown). This will be considered further in chapter "Valuing Ecosystems".

Cultural Services

Cultural services is an extremely broad component of Ecosystem Services. This breadth of definition becomes particularly apparent when attempts are made to apply valuations to them. When describing Cultural Services, TEEB identifies a range of different services.

The use of the natural environment as a source of recreation and health not only includes natural ecosystems, but also urban greenspace and other open areas. It is not dependent on the ecosystem being functional, but rather needs to be physically accessible. This provision is therefore defined primarily through the geographical constraints of proximity and accessibility. One element of this use of the natural environment is that it provides sustainable benefits to the users only up to a particular level of activity. Highly frequented areas have a tendency to suffer degradation through litter, noise, erosion and increased disturbance. This leads to a decline both in the benefits appreciated by the user, but also reductions in the ecological health of the environment itself. This paradox raises significant policy issues, as if an area is open for the free use of all it runs the risk of losing its ability to provide the service it has been selected for. If there is an optimal level of activity which is sustainable in the long term how is this to be determined? More controversially, how is this to be enforced?

The value of ecosystems and biodiversity in the context of tourism is identified as another cultural service. This has the ability to offer economic benefits to host areas. It represents a logical progression from the benefits of local green spaces, where individuals travel to other areas for what is perceived to be more interesting environments. Of course, the issues relating to the overuse of green spaces also apply to popular tourist locations, but with an economic counterbalance in that more tourism results in more revenue (Fig. 3).

From an economic perspective, it is important to bear in mind that revenues received from tourism are not secure and are not in the control of the country or region where tourism represents a significant income stream. Any reduction in spending power within the more developed countries which supply the majority of tourists will result in a decline in tourism revenues, as individuals cut back on luxuries such as holidays and are more frugal in their spending habits.

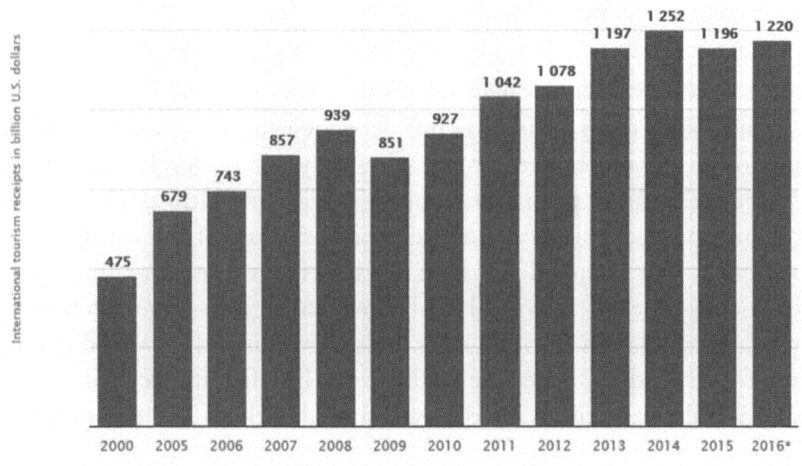

Fig. 3 Global tourism revenue (*Source* statistica.com)

If a decline in tourism does occur there is the danger that the natural attractions which were formerly maintained and acted as a source of revenue will be either neglected or converted to more profitable land uses.

Two additional cultural services which are presented are the aesthetic and spiritual appreciation of the natural environment.

The valuation of cultural services represents a significant hurdle in the development of any comprehensive valuation methodology. Some aspects such as the income received from tourism is a relatively straightforward calculation. If health benefits could be quantified then these too could be valued in terms of a reduction in costs of healthcare.

However, aesthetic and spiritual values represent very different challenges. Aesthetic value is quite literally in the eye of the beholder, and spiritual value lies in the heart of the believer. There is no means by which any external economic mechanism can be applied, other than the provision of some right of ownership to those who are most deeply affected. This has occurred in some cases, where public fund-raising activities have allowed a particular site to be purchased, but in cases where those who are most intimately associated with a particular feature do not have sufficient economic leverage to enable free market forces to operate, it is impossible to apply a value from a position of impartiality.

It is important to note that biodiversity itself is not considered to be an Ecosystem Service, instead it is described as that which underpins the supply of services. The value of biodiversity within TEEB is considered to be captured under a cultural Ecosystem Service called 'ethical values'. The introduction of an ethical or moral standpoint presents a new range of difficulties when considering valuations within the economic sphere. This is particularly acute when considering such a poorly understood concept as biodiversity.

For example, the practice of bloodletting in medicine was based upon incomplete knowledge and inaccurate models of human health. Although this practice had no positive health benefits, and could in fact be harmful, there is no question that the practitioners were acting in a moral or ethical manner. So to present a particular course of action as ethically or morally valuable in the face of a deficit of knowledge can present a wide range of unintended consequences.

Discussion

The outline of the classification of Ecosystem Services presented here clearly shows that a single mechanism of valuation cannot capture all of the potential services provided by the natural environment. Not least because many such services remain unknown and unrecognised in our present state of understanding.

The possibility of substituting many Ecosystem Services, such as engineered flood relief schemes and the practice of reducing a broad ecosystem to those elements which are of greatest benefit to mankind through selection of ecosystem components for domestication begs the question of how essential are ecosystems in their natural state to the provision of the services which are obtained from them. This presents a significant policy question in relation to the relevance of Ecosystem Services to the wholesale value of an ecosystem.

Ecosystem services are, almost by definition, a reductionist approach to the classification and categorisation of parts of an ecosystem which are of value to mankind. However, there remains a significant element of the ecosystem approach which we are ignorant of. There are many cases where we simply do not have the information to make an informed decision. For example, it is impossible to know how many potential medicinal chemicals occur in a natural system.

This means that purely economic valuation has to be focused on those elements which are known and have an identifiable value, either as a prevailing commercial market in natural goods, a value representing substitute costs or the value of extracting such elements from the ecosystem.

Despite these issues, one of the main aims of TEEB was the creation of a system which enabled the valuation of Ecosystem Services. The approach adopted the following three core principles.

Firstly, the recognition by local communities or wider society of some form of non-monetary value in ecosystems, landscapes, species and other aspects of biodiversity. This form of valuation is more accurately considered to be a recognition of cultural importance, and is closely associated with belief systems and the social structure prevalent in any given community. TEEB considers this to be sometimes sufficient to ensure conservation and sustainable use. Although this aspect is not something that can be integrated into an economic model, the control of the area considered to be important, through ownership of the land, rights of access and the resources required to maintain its importance can certainly be incorporated into the economic sphere.

The second principle adopted is the demonstration of value in economic terms to policy makers and business, to allow them to undertake a cost/benefit analysis which takes into account Ecosystem Services. Of course, this demonstration of value can only take into account the known elements of any ecosystem which provide a currently recognised service. This does not include any value present in an ecosystem which has not yet been recognised. It requires an ecosystem to be considered a 'black box' which provides an economic service, as knowledge of the details of such systems are unavailable.

Finally, TEEB concerns itself with capturing value. This principle promotes the use of economic mechanisms, such as incentives, subsidies and taxation to achieve what is perceived as environmentally positive outcomes, based upon existing knowledge. This approach epitomises the mainstream economic model of applying fiscal policies in order to achieve theoretically beneficial outcomes.

The TEEB approach to valuing Ecosystem Services is from a hierarchical point of view. The principle of this approach is that elements of an ecosystem which provide services can be valued and therefore protected through policy in a top-down approach. Through the application of such policies, other non-service providing elements of the ecosystem would also be subject to the same policy decisions and protected as an incidental by-product. The subject of valuing Ecosystem Services is considered in more detail in chapter "Valuing Ecosystems".

Mainstream Economics

For many people, economics is very much seen as something remote and 'the way it is' in terms of human life. It is considered to be the domain of banks and governments. The degree of economic freedom an individual has is based upon the level of pay, the tax regime, interest rates to be paid on loans and the prices paid for available items and services—ranging from the essentials of life to luxury goods. From the perspective of individuals, all of these factors are determined by outside influences.

This section will introduce the basic theory under which the global economy currently operates, which can be described as a combination of Keynesian and neoclassical economic theory, commonly known as 'Mainstream Economics'.

Mainstream economics believes that it is possible to model the economy mathematically and identify interventions and modifications to the system, implemented through changes in policy affecting such things as interest rates and levels of taxation, from which the state, or other tax raising entity, can provide communal services and invest in a variety of other projects. The aims of such interventions are to achieve growth in the economy (normally measured as an increase in GDP, or a similar metric, Fig. 1 in chapter 1 "Introduction"). It is believed that this economic growth has the direct causal effect of allowing individuals to maximise their economic objectives (which is to obtain and use goods and services they find the most beneficial, i.e. having the greatest 'utility'), whilst simultaneously limiting the constraints on achieving these aims (due to income, prices or resource availability).

This approach contains a number of assumptions which are taken as received wisdom by the majority of politicians, economists, investors and other operators within the financial sector.

At the level of the individual consumer or business entity, mainstream economics considers that all decisions made are rationally based upon the common goal of maximising their economic objectives (improved standard of living, profit). It is only by adopting this view, that all decisions have a rational and commonly based foundation, that is remotely possible to formulate and apply modelling techniques.

Mainstream economic theory also adheres to the belief that economic interventions by governments are the way to achieve these goals. It sees the 'economy' as a mechanistic process, which requires an overview of its component parts and has an inherent need to be managed or controlled by state policy and procedures. In any given situation it is believed that interventions are required to correct the economic mechanism in order for it to operate as it is supposed to do.

This interventionist approach to the economy has strong parallels with the concepts of managed environments, where it is seen as a necessity to make alterations to the habitats and ecosystems considered to be of importance, in order for them to meet an ideal view of a healthy environment.

The foundation of modern mainstream economics can be found in *The General Theory of Employment, Interest and Money* by John Maynard Keynes [11], which establishes a series of key economic theories.

Keynes identified one of the key requirements of a well-functioning (i.e. growing) economy as high employment, which results in wages increasing due to a limited supply of labour in the marketplace, which can therefore demand higher levels of remuneration. This, in turn, results in an increase in spending by wage earners as they have a higher level of 'disposable income'. The increase in expenditure acts to further stimulate the economy through increased production of the consumer items the workers desire and so more workers are needed, further reducing unemployment and continuing the pressure of increased wages. One effect of the increase in wages is an increase in prices for goods (as the costs of production increases due to higher wages). Keynes saw this type of inflation as a sign that an economy was functioning correctly.

According to theory, when this virtuous cycle of increasing employment, rising wages and consumer activity, breaks down then the situation goes into reverse. When this occurs there is an increase in unemployment and less purchasing activity. This decline in spending then leads to declines in activity in other parts of the economy.

The solution to this situation identified by Keynes was for government intervention. Such intervention was considered to be a necessity, as the health of the economy was the basis of government income (through taxation). Therefore, the state as an institution has a vested interest in maintaining growth.

In order to provide a return to the cycle of growth, the state would borrow money (through the issuance of government bonds or other financial vehicles) and use it to provide employment for its citizens (Fig. 4). This, according to theory would, in turn, lead to more spending and a return to a cycle of growth.

At this point, it should be noted that it is the increase of numbers of people obtaining an income which is the primary aim of such Government spending. The theory does not require the employment to actually result in the production of goods and services which have utility, as these items would be generated within other sectors of the economy as a result of an increase in spending power of the state employees. It is also worthy of note that the debt incurred by the government is in fact a debt borne by all the citizens of the state.

One side effect of this view of economics is that individuals are seen as little more than a conduit of money. They receive wages which are then spent. In order for this to operate successfully, the amount of money retained by individuals (via savings and investments) should be limited as far as possible.

The use of inflation as an indicator of economic health has this very effect on the consumer, that is the purchasing power of any given currency will decline over time. For example, in order to purchase the equivalent of £100.00 of goods in 1971, a worker in 2018 would need to have £1344.00 (measuringworth.com). This does not indicate that we are working any harder or being more productive than we were in

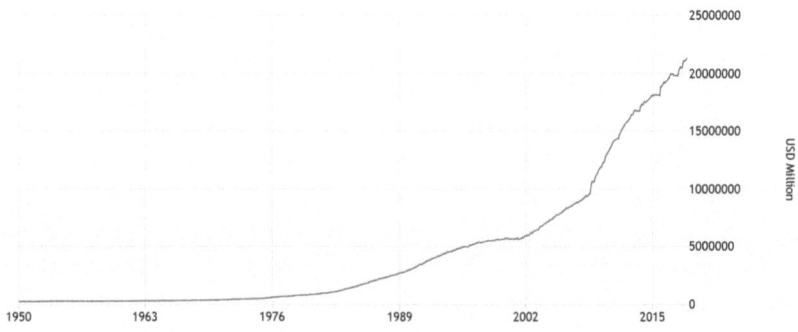

Fig. 4 US government debt (*Source* Tradingeconomics.com | US Department of the Treasury)

1971, all it means is that there is much more money in circulation, and that by keeping hold of money (saving) over time there is a loss in purchasing power. How this extra money came into existence will be discussed shortly.

In the current view of mainstream economics, the concept of government intervention does not only apply to individuals, but also at the level of businesses. Governments view a business as an 'economic unit' which behaves in a similar fashion to an individual in the above explanation. When the economy is doing well, nations businesses are also doing well and making profits which benefit investors (through higher share prices and payment of dividends), this benefits a nations GDP. However, when businesses perform less well there is a decline in share value and less wealth in the economy. Under such circumstances, the state (via Central Banks) can effectively lend money to businesses through the purchase of corporate bonds, providing them with the cash required to continue to grow.

Keynes did not criticise governments for having an unbalanced budget, that is spending more than their income. The theory promotes the concept of policies which are designed to act against the direction of the business cycle. The intention of this is to reduce the economic fluctuations which are observed to produce an economy which is as stable and predictable as possible.

It is important to place these theories in historical context. They were formulated at a point when western industrialised civilisation had just endured a major economic downturn, so the concept of stability was in the forefront of the minds of governments and economists. The economy at that time was becoming increasingly dominated by corporate enterprises which were involved in labour-intensive enterprises. Employment was becoming highly centralised with corporate interests employing vast numbers of workers, the employer/employee model was the main platform from which economic theories were derived. This, alongside a belief in increasing control by benevolent centralised government, naturally lead to an adherence to the type of economic theories formulated by Keynes.

Keynesian economics dominated economic theory and policy until the 1970s. At this point, the phenomenon of 'stagflation' (a combination of high inflation and slow growth) became widespread in many advanced economies. This phenomenon began to raise questions regarding the ability of governments to regulate the business cycle with fiscal policy (government borrowing and investment). An alternative viewpoint began to increase in popularity, that Central Banks had a greater role to play. By controlling the supply of money in the system and manipulating interest rates it would be possible to alleviate the situation. Increasing the amount of money in the system would serve to increase inflation in the short term, but the overall effect in the longer term would be increased economic activity, leading to growth.

The combination of fiscal policies from government and the monetary policies of Central Banks were then both adopted into the mainstream. This two-pronged approach to managing the economy prevails to the present day.

It may be surprising to appreciate that the current system of economics, which is dominated by the overarching need for growth, is in fact not the only way in which economic activity can be viewed. There exist many schools of thought which promote a variety of different approaches to the economy and there is no proof that the theories which drive the economic system that we all live under are optimal for the majority of people. This is reminiscent of Churchills quote regarding the adoption of a democratic political system from the introduction to this book. In this case, however, there is no evidence that the prevailing economic system is in fact the 'least worst' from the alternatives available.

Critical Review of the Prevailing Economic Model

Money

Valuation is a fundamentally economic activity, within which there is a need to define the 'price'. This leads us into the realm of considering how price is set in the economy, how it is measured (in what denomination do we establish a price) and its variation over time (against what benchmark is it set). More fundamentally, it requires us to understand the concept of money, how it is created and used.

Valuing Ecosystem Services is inevitably framed in monetary terms. Projects and developments which have the potential to either deplete or enhance Ecosystem Services require some form of monetary resource to plan and implement them, be they governmental or other public sector investments, or the result of private sector corporate planning. This section considers the nature of money, its universal characteristics and how it is brought into existence.

Money in its original form represented a store of wealth, received through the expenditure of time and effort in productive activity. This store of wealth could then be exchanged for goods and services when it is translated into an exchangeable form, generally this is some form of currency. This leads to an important distinction between currency (the means of exchange) and money (the store of wealth). In this scenario, wealth is created through work and once created cannot be removed from the system.

This situation assumes that the value of such wealth is stable over time and retains recognition as such through the medium of money. Of course, for this situation to hold true it is necessary for money to be in some way associated with a stable foundation. Traditionally this foundation has been based upon precious metals. Any currency would be defined in terms of its relationship to a certain amount and purity of a precious metal (usually, but not exclusively gold and/or silver). For example, the *US Dollar* is a specific coin defined in the United States Constitution as containing 371.25 grains (troy) of fine silver. In Anglo-Saxon England, the most common coin in use was the silver penny. 240 pennies, which weighed one pound, was the origin of the modern British currency, the Pound Sterling. Sterling silver is a purity of 92.5%, so the original Pound Sterling was nothing more than a weight and designated purity of precious metal.

This form of system, where a currency is established against a fixed precious metal standard, offers a number of advantages to ordinary citizens, but limits the economic powers of governments.

Citizens benefit in the following ways:

- Income is either composed of (via coinage) or convertible into (via a promissory note) a standardised quantity of precious metal which cannot be devalued by government decree.
- The exchange rates between currencies on the same precious metal standard are fixed. In fact, coins of a known weight and purity of metal are universally acceptable, irrespective of the minting nation.
- Money received for work undertaken, in terms of coinage, does not require any belief in the ability of a third party to meet its obligations. It allows a transaction to be completed through a material transfer.

Disadvantages for government include:

- Debt cannot be reduced through intentional devaluation of a currency.
- Speculative borrowing requires more stringent oversight, as repayment cannot take account of inflationary devaluation of a currency.
- Mainstream economic principles of economic stimulation through the creation of additional money supply cannot be undertaken without also increasing productivity.

Throughout history, the disadvantages of a precious metal based monetary system have lead states to produce currencies which are made legal tender by fiat (decree). Such currencies are of no value in terms of their materials (paper and base metals) and are not convertible into precious metals at a fixed rate. The value of such fiat currencies is based upon confidence that its value will be honoured between the holder of the currency and the owner of the goods and services they desire to purchase. The currency supply is out of the hands of the owner of currency, therefore the amount in existence and its commensurate purchasing power is controlled by the currency issuing institutions. All currencies in circulation today are fiat.

Therefore, by controlling the quantity of money in circulation, issuing institutions play a controlling role in the wealth of individuals through time who are subject to transact in that currency, as their wealth can be reduced through the increase in supply of the tokens which represent their accumulated wealth.

This is a tool used by political and financial governance to provide a stable economy as increasing the money supply is considered to be a method of stimulating growth in an economy (by encouraging increases in spending). However, the downside of this approach is that it effectively devalues any wealth which has already been stored (saved).

In effect, this system forces individual economic actors (be they individuals or businesses) to actively engage any accumulated wealth into the financial system, as this is the only means by which wealth can be preserved. This means that savings are actively discouraged, so personal or collective insurance against times of economic downturns (when normally savings would be spent) is subsumed into a dependence on overarching economic control to provide the necessary stability.

Not only does this have the effect of causing an entrained social dependence upon a single control system, but the speculative nature of investments rather than savings can lead to an exaggerated decline in collective wealth when a downturn occurs in the economy.

By devaluing savings, their use as independent capital for future economic ventures is discouraged. Rather, they are thrown into a collectivised pot where access is prioritised for larger, existing enterprises. This has the effect of promoting the development of large, publicly listed corporations rather than small independent businesses.

In order to embark on a venture, it is unusual to be able to use individual savings to develop a business. Borrowing from the financial system is not only essential, but is encouraged at all levels.

Fractional Reserve Banking

The use of precious metals as money led to logistical issues in relation to transportation of wealth and its vulnerability to theft. For safety, money over and above that required for day to day use, was deposited with a secure institution (often a goldsmith) for safekeeping. In return, a paper receipt was provided to the depositor.

This receipt would allow the deposit to be redeemed, but the metals deposited did not have to be the same as those returned. Only the weight and quality of the metal was of concern. This procedure was adopted by the Knights Templar, who would receive deposits of gold in Europe from pilgrims. They would be provided with a paper receipt which could be redeemed from Templar holdings in the Holy Land. This meant that there was a lesser chance of robbery during the pilgrimage, especially once it became clear that the pilgrims were only carrying apparently worthless pieces of paper rather than immediately accessible valuable assets.

Eventually, as the practice spread, the paper receipts could be exchanged between depositors in good faith that the underlying precious metals were securely held. In this way, the use of paper currency and the rise of banks were established.

Banking institutions soon realised that the precious metals held on deposit were unlikely to be all required by depositors at the same time. From this, it became clear that banks could produce more printed receipts (currency) than they had underlying assets on deposit. These receipts could then be lent to borrowers, who would be required to pay back the banks, with the addition of supplementary interest. In this way, paper currency was actually created by the demand from borrowers, rather than originating from productive labour. This situation remains in place to this day. When a bank makes a loan to a borrower, the money is then brought into existence, to be paid back (with interest) over time. Where the additional money for the interest comes from in such a closed system is generally not discussed!

Through the use of this mechanism, banks could profit directly from being a trusted depository of precious metals and having paper currency which would be widely accepted as a commercial substitute for tangible precious metals.

This system (known as fractional reserve banking) remains the foundation of the financing of loans to this day. One key side effect of this system is that in order to create profit, banks must lend money. In the days of the gold standard, the amount of precious metal limited the ability to loan. This is not now the case.

Management of National Economies

In mainstream economics, the collective economy is usually operated in a two-handed manner. Firstly, there is the role of the government, which involves the generation of revenue through taxes, sales of government bonds, etc., and the expenditure of this revenue to maintain a stable economy and promote other political aims (fiscal policy). Secondly, the central bank controls the supply of money in the economy and sets interest rates appropriate to the prevailing economic environment and in accordance with economic theory (monetary policy).

The targets of a national economy are based around growth (as indicated by growth in GDP) and inflation, which is thought to be a reflection of economic health. In order to achieve these aims, there is an overwhelming need for commercial activity, which frequently involves land use changes such as mineral extraction, development for housing, industrial expansion and construction or improvement of infrastructure such as roads, railways, etc.

Such activity is paid for either through government projects (via fiscal policy) or through private corporations which raise capital through the banking system. The extent of borrowing is determined by interest rates and availability of loans, which are dictated by monetary policy.

It is very rare for projects to be initiated without the need to raise capital against future growth or profits. Because most projects are founded upon the need for debt (which is welcomed by financial institutions, as they receive interest on any loans made) almost any imposition of economic costs at the beginning of a project is not only easily capable of being subsumed into future economic projections, it also represents a benefit to financial institutions who will receive more interest on the larger financing requirements.

As the environmental cost of a project occurs at its instigation it is underrepresented in terms of overall costs into the future (which increase as a result of year upon year compound inflation, a requirement of growth). In addition, even if the project is a failure or at least falls short of any projected economic benefits, the loss of environmental components has already occurred.

Thus, the mainstream economic system where growth and inflation are necessities against the background of the potential for infinite fiat money creation makes it extremely difficult to apply the mechanisms of Ecosystem Service valuation in a manner which would lead to cost-benefit assessments resulting in the costs environmental loss outweighing prospects for economic growth.

Mainstream Economics and Ecosystem Services

The concept of Ecosystem Services, as a mechanism by which aspects of the environment may be valued. is based very firmly within the worldview of mainstream economic theory, including a range of presuppositions and assumptions. This section highlights the various aspects of the Ecosystem Services approach which are inextricably linked with mainstream economic theory.

Economics Is Driven by State Policy

One of the key purposes of TEEB and its elaboration of an Ecosystem Services approach is to help inform government and policy. There is therefore an explicit focus within the framing of Ecosystem Services of a top-down approach to economics as it relates to the environment.

It is assumed that by providing information to the top of the decision-making hierarchy there will be a response via legislative and policy instruments which will result in appropriate outcomes. This is fundamentally the *modus operandi* of all mainstream economic theory which gives primacy in economic action to the state and the central banks operating in concert to drive the economy in a predetermined direction. Through adopting the status quo, Ecosystem Services can be seen as in fact offering a gateway for the extension of mainstream economics into the environmental arena. This means that the economy and the environment would both be subject to increased control and regulation both through statute and economic mechanisms such as subsidies and taxes, with the intention of reaching a theoretically determined state of 'sustainability'.

That Economic Growth Is Both Necessary and Advantageous

In the mainstream, economic growth (as measured via GDP) is established as the only means of improving standards of living and simultaneously reducing debts. Such growth is essentially a function of the economic system, rather than the economic activity driving growth. Government policy is designed to entrain economic activity towards a scenario of constant aggregate growth, affecting the economic behaviour of individuals and businesses in the process.

The formulation of Ecosystem Services is one which embraces the idea that such growth is a necessity. Rather than questioning the need for such growth leading to environmental damage, it is framed in a way which provides answers to the question of how much additional cost the damage of (inevitable) growth to ecosystems will be.

In this respect it fails to address the significant question of how growth is financed and whether it is more important to be productive or merely engaged in activity (which is a distinction that is poorly defined in mainstream economic thought). An emphasis on state-funded activity as a means of economic engineering has the potential to lead to large scale interference of natural environments and processes in the name of conservation or environmental management which, although instigated with the best of intentions, has the potential to cause significant damage, particularly to the elements of biodiversity which we currently have no knowledge of. By focussing on the known elements of an ecosystem we could inadvertently adversely affect the unknown elements. But who would know?

The Need for Management

Economics has moved itself as a subject away from the actions of individuals towards overarching theories which have increasingly considered the economy as something which is to be managed and controlled by relatively few individuals. Ecosystem Services is presented as a tool to enable sensitive management of environmental issues. This approach presupposes that there is a fundamental need for the environment to be managed through policy and legislation. Although in many cases this is a genuine need, the reasoning behind this is not fully expressed.

In fact, such interventions are an artefact of the current operation of mainstream economics, where it is activity, not productivity which is the primary measure of growth. This means that even though an action may be environmentally destructive it is considered viable and 'without an economic alternative'. This is purely an artefact of the prevailing economic paradigm, and should be understood as such.

Projects for conservation and legislative protection affect the operation of an ecosystem. These are perceived as wholly beneficial and are worthy uses of the financial income derived from the valuation of Ecosystem Services. Notwithstanding the grey area regarding the sources of the financial income, the belief that intervention is of benefit in a self-organising system (which is recognised in ecosystems far more acutely than in the world of economics) is something which is open to considerable debate in both spheres.

The Use of Modelling

Both ecology and mainstream economics have adopted the use of mathematical modelling as a means of providing a means of extending understanding of a system and allowing transference of observed phenomena in one particular case to a wider and more generalised principle.

However, the use of such an approach requires certain precautions to be taken. In order for the model to be used in a new situation (in ecological models this could relate to either a different location or at the same location during a different time period) there must be a reasoned understanding of why the model is applicable under different circumstances. Without this rationale, it is possible to develop guidance on habitats which are at best suboptimal and at worst counterproductive. In addition, the production of a broadly applicable model inevitably results in a loss of resolution. As it is not possible to model the activity of every economic actor or of every individual in a species population, there is a need to use metrics and aggregate trends. When this approach is adopted in the development of valuation models for Ecosystem Services, it is all too easy to lose sight of the cost of a loss to an individual economic actor who is dependent on any given service for their livelihood and well-being.

This view of decision-making being out of the hands of the individual actors with a direct involvement in the environment reflects the mainstream economic view that it is possible (and desirable) to manage individual choices through a top-down hierarchy which has a greater vision of what is required and can provide appropriate guidance.

This implicit control system reflects the notion that the use of modelling and the concept of an ideal system can be transferred from the economic activity of individuals to the management of the environment. In addition, the belief that the empiricism of mainstream economics can be used to develop policy includes an assumption that all 'economic actors' will respond in a way which is in accordance with the modelling used to develop the policy. This is a particular difficulty in situations where the model used is based upon only partial information and limited capacity for empirical testing.

By working within the constraints of the existing economic model, the concept of Ecosystem Services loses its ability to question the validity of mainstream economics as a means of valuing the natural environment.

Austrian School of Economics

Even though the ideas of Keynes and other aspects of mainstream economics are almost universally adopted by governments worldwide, there are several schools of economic thought which diverge from its basic principles in significant ways. One such alternative view of economic reality is the 'Austrian School' of economics.

Whilst mainstream economists have adopted mathematical modelling and a simplified aggregate approach to the behaviour of individual economic actors in their attempts to predict the future of the economy, the Austrian School is based upon the use of logic and reasoning to understand human economic behaviour and processes. It follows the concept that social phenomena (including economic activity) result from the motivations and actions of individuals, based upon each having a unique view. They believe that an aggregate view of the economic behaviour of a population has no predictive value and that the explanations presented for previous economic actions do not necessarily follow any set model or theory.

This approach, known as methodological individualism, originated in late nineteenth and early twentieth century Vienna with the work of Carl Menger, Eugen Böhm von Bawerk, Friedrich von Wieser and others. Current-day economists working in this tradition are located in many different countries, but their work is still referred to as Austrian economics.

The Austrian School theorises that the subjective choices of individuals including individual knowledge, time, expectation and other subjective factors, cause all economic phenomena. Therefore, Austrians seek to understand the economy by examining the social ramifications of individual choice. In this fundamental respect, it differs from other schools of economic thought, which have focused on aggregate variables, equilibrium analysis, and societal groups rather than individuals.

In 1949, Ludwig von Mises presented a version of this approach, which he called 'praxeology', in a book published in English as Human Action [12]. In it, Mises argued that praxeology could be used through the means of undertaking deductive economic thought experiments which would yield logically derived conclusions. He wrote that conclusions could not be inferred from empirical observation or statistical analysis and argued against the use of probabilities in economic models.

To provide an ecological analogy, the contrast between mainstream and Austrian economics can be seen through the perspective of how one views the process of natural selection. It is perfectly possible to identify the environmental features which lead any species to have developed the characteristics which it currently has, based upon the principle that over many generations its survivability has been optimised. However, it is not possible to make any kind of prediction of future evolution on this basis. As an example, because multicellular organisms are very common, it does not follow that the path of evolution for a present-day amoeba will be to develop a tendency towards multicellular life.

Although the principles of natural selection apply throughout time, as do the principles of economic activity for humanity, they have limited ability to predict future direction. The idea of humanity undertaking conscious management of evolution towards some theoretically idealised future species is certainly in the realm of horror fantasy fiction, but the aim of engineering our future economic actions, based upon the events of the past, is accepted by all.

In order to provide an explanation of economic phenomena, the Austrian School investigates the actions of individuals, based upon their unique judgements and choices. It emphasises that such choices are based upon whatever knowledge they have or believe to have and whatever expectations they hold regarding external developments. In particular, the likely consequences of their actions are given particular prominence.

A summary of the differences in approach between Austrian and mainstream economic worldviews are presented in Table 1.

In mainstream economics, the business cycle is generally defined according to the work of Arthur F. Burns and Wesley C. Mitchell [13] in their book *Measuring Business Cycles* (1946):

> Business cycles are a type of fluctuation found in the aggregate economic activity of nations that organize their work mainly in business enterprises: a cycle consists of expansions occurring at about the same time in many economic activities, followed by similarly general recessions, contractions, and revivals which merge into the expansion phase of the next cycle.

Under mainstream economic theory, this cycle is controlled by state interventions which attempt to reduce the adverse effects of contraction on citizens, and making the economy grow as smoothly as possible. The industrialised economies are seen as a system of interrelated components, organised as a network of free enterprises searching for profit and all operating under the same assumptions. This means that according to theory they will all act in a predictable manner, following the logic modelled by mainstream economics.

Table 1 Austrian and mainstream economic worldviews

Topic	Austrian approach	Mainstream approach
Trade	Free trade, in which there are no state barriers or subsidies to conducting commercial trade activities. Willing purchasers and sellers should have the freedom to interact and conduct business without interference	Trade needs to be controlled by government to maintain domestic growth, protect the domestic workforce and companies, through mechanisms such as subsidies, trade tariffs and taxation
Currency	Money should be based upon some form of commodity standard, such as precious metals. The ability to create money should be strictly controlled and limited, it should retain its value over extended periods of time. Money is created through productive activity	Money is flexible and fiat in nature. The quantity of money available is authorised by the state, to be used as a tool in maintaining economic stability and mandated growth. Money is created through the establishment of loans
Growth	Economic growth occurs naturally, through the use of savings from profitable activities to invest as capital in future ventures	Economic growth is mandated and occurs through the generation of money to create economic activity and increased consumption. Future ventures are funded through borrowing on the basis of projected profits
Inflation	Inflation represents a devaluation of money and savings, it is a sign that the economy is not balanced	The state establishes a level of inflation which should be maintained to ensure continued spending and consumption (if an item will be more expensive in the future it is better to purchase it immediately). Maintenance of spending is a sign of economic health
Support	All businesses and corporations which find themselves in financial difficulty should be allowed to fail without external influence	Larger businesses which have a disproportionate effect on employment and growth should be supported by the state through loans and subsidies, in order to prevent their failure at all costs

In contrast to this view, Austrian economic theory views business cycles as the result of excessive bank lending, due to artificially low interest rates. This increase in lending results in an imbalance between saving and investment.

In the Austrian view, the business cycle comprises a series of actions as follows. To stimulate the economy, interest rates are lowered, which both reduces the amount of savings in the economy (as lower interest means less reward for savers) and increases the amount of borrowing from the banking system (credit is cheap to repay). The effect of this is an increase in spending, which is funded by newly issued bank credit.

This does have the desired effect of stimulating the economy in the short term. However, the cheapness of credit means that much of it is spent on activities which are not, nor can ever become, profitable. This 'malinvestment' eventually becomes visible in the economy further down the line, when unsustainable businesses are even unable to repay the (low) interest rates which were used to borrow the capital in the first place. Once this lack of profitability becomes recognised, lenders realise that they are unlikely to be paid back and so limit the amount of lending they carry out to new borrowers. Furthermore, they lend at significant higher rates, in order to try and cover previous losses. This, in turn, reduces demand for credit and economic activity declines, leading to a recessionary phase.

Austrian economics views this as a natural cycle and an inevitable result of humans acting in a rational way in relation to the prevailing economic climate. State intervention in this cycle is generally viewed as counterproductive to the maintenance of a healthy economy.

Where Ecosystem Services Can and Cannot Adapt to Alternative Economic Models

This discussion aims to demonstrate the ways in which Ecosystems Services theory would have to adapt in the event that mainstream economics shifted, or was even superseded by an alternative economic worldview.

How Ecosystem Services Was Developed in a 'Managed Economy' and Features of It Which Are Dependent on This Model

The purpose of Ecosystem Service valuation is to act as a financial demonstration of the importance of Ecosystem Services to policy makers, with the aim of providing a framework by which decisions relating to potential damage to ecosystems, and their inherent 'services' and biodiversity can be evaluated in fiscal terms, as part of an overall cost/benefit analysis.

At a fundamental level, the introduction of Ecosystem Services into policy- and decision-making activity represents a trade-off between the cost of protection and maintenance vs the cost of destruction. In that respect Ecosystem Services represent a non-exploitable resource and a standing cost. However, in a debt based economy which is predicated on national and corporate growth as a prerequisite to service interest payments on debt the cost of third-party destruction can always be met by increasing debt.

In accordance with mainstream economics, the system of Ecosystem Service valuation presents an aggregate view of value. It employs the mainstream definition of money as a means of exchange. This valuation therefore implicitly defines Ecosystem Services as a tradeable commodity. This is in stark contrast to the valuation of Ecosystem Services as an asset.

As a commodity, an Ecosystem Services value will fluctuate in relation to the availability of credit. At times of low interest rates and easy credit availability, it is more likely that a cost-benefit analysis would show the loss of Ecosystem Services to be an acceptable sacrifice. Where money is more expensive, an Ecosystem Service value is likely to be considered to be a more significant component.

If Ecosystem Services were considered to be an asset then their value would remain stable or change in a countercyclical manner. For example, if a state brought 30% more money into circulation in order to stimulate the economy then the value of Ecosystem Services would increase to a fixed proportion of the overall money supply (in this case by 30%).

Ecosystem Services, as currently framed, implicitly acknowledges that the current economic system based upon the need to service debt obligations, such as interest payments and shareholder dividends, requires a level of economic growth which inevitably leads to increasing demands upon natural resources and the intensification of production. The response of the Ecosystem Services approach is to establish a system of prioritised areas to be lost, rather than address the failure of the underlying system.

It operates on the principle of a hierarchical system of control in terms of policy- and decision-making. This approach does not provide for action or ownership of natural resources at a dependent community level. By identifying Ecosystem Services value as a single metric which can be used as part of an analytical model and incorporated into decision-making, it fails to deal adequately with the utility value of Ecosystem Services to a range of stakeholders, each of whom would value such services differently.

Praxeology, the Austrian school study of human choice and actions, takes an individualistic and narrative approach to economics rather than an analytical and quantitative approach. Such an approach, when applied to Ecosystem Services is more likely to present outcomes which benefit the immediate users of any given Ecosystem Service rather than the good of a state or nation. Especially when characterised in terms of national economic growth.

However, there is a similarity between the approach of Austrian Economics in terms of praxeology and the consideration of the role of humans in the wider environment in respect of Ecosystem Services. The key point is that economic development and the well-being of humanity must of necessity be considered at the individual level. Well-being can only be measured individually, rather than in aggregate.

The Universal Thread Which Is Relevant to All Models

The most important aspect of the development of Ecosystem Services is that it shines a spotlight on the fact that the benefit humanity gains from natural systems is not generally taken into account in economic theory, be that mainstream, Austrian or otherwise. The systematic identification and recognition of Ecosystem Services provides a mechanism which could significantly affect decision-making in the economic sphere.

However, it must also be recognised that there are large gaps in current knowledge regarding how ecosystems operate and what they consist of, both in terms of overall biodiversity, and also in the arena of unrecognised potential for future Ecosystem Service provision.

This raises the important issue of a knowledge deficit in terms of Ecosystem Services and biodiversity studies. If decisions are based only on our existing knowledge and current understanding, how can there be any confidence that humanity, as a unit, is making the correct decisions regarding biodiversity when we only have knowledge of the existence of approximately 10% of the subject of our actions?

In the MEA, the concept of intrinsic value was used as a mechanism to rationalise activity to maintain ecosystems. Rather than invoke this concept of 'intrinsic value' it would be more appropriate to allow for such actions to be works based upon a precautionary valuation for unknown or as yet unused utility. This allows for a value to be placed upon the unknown rather than ascribe a purely altruistic motivation which is rarely realised. It therefore allows the ecosystem to enter the economic sphere as a reservoir of potential.

This raises the priority of filling this knowledge gap to the highest level, as it is only by understanding the specifics, rather than projected aggregates, or broad assumptions based on a limited understanding of the system, that we will be able to make decisions based upon fact rather than speculation.

In economic terms, the expenditure required to fill the knowledge gap is not inconsiderable, and difficult to quantify. The issue is that the work required is not productive in the sense of providing economic benefits, but rather is a tool by which decisions can be made with a degree of clarity in the future.

Rather than seeking in the dark for a partially formed concept of 'the value of nature', it is conceivable that economic methods could be employed to perceive 'nature as value' in a way which would financialise the unknown and allow human effort to be applied to bringing it into the realm of knowledge.

References

1. Kasler, D., Wilke, C., & Sabalow, R. (2017). Endangered Frog Habitat Sparks California Farm Lawsuit. *The Sacramento Bee*.
2. Bell, M. L., Davis, D. L., & Fletcher, T. (2004, January). A Retrospective Assessment of Mortality from the London Smog Episode of 1952: The Role of Influenza and Pollution. *Environmental Health Perspectives, 112*(1), 6–8.
3. Lovelock, J. E., & Margulis, L. (1974). Atmospheric Homeostasis by and for the Biosphere: The Gaia Hypothesis. *Tellus, 26*(1–2), 2–10.
4. Millennium Ecosystem Assessment. (2005). *Ecosystems and Human Well-Being: Synthesis*. Washington, DC: Island Press.
5. Holdridge, L. R. (1947). Determination of World Plant Formations from Simple Climatic Data. *Science, 105*(2727), 367–368.
6. www.teebweb.org.
7. TEEB. (2008). *The Economics of Ecosystems and Biodiversity: An Interim Report*.
8. TEEB. (2010). *The Economics of Ecosystems and Biodiversity: Mainstreaming the Economics of Nature—A Synthesis of the Approach, Conclusions and Recommendations of TEEB*.
9. TEEB. (2010). *The Economics of Ecosystems and Biodiversity Ecological and Economic Foundations* (P. Kumar, Ed.). London and Washington, DC: Earthscan.
10. TEEB. (2011). *The Economics of Ecosystems and Biodiversity in National and International Policy Making* (P. Ten Brink, Ed.). London and Washington, DC: Earthscan.
11. Keynes, J. M. (1936). *The General Theory of Employment, Interest and Money*. Cham: Palgrave Macmillan.
12. Mises, L. von. (1949). *Human Action*. New Haven: Yale University Press.
13. Burns, A. F., & Mitchell, W. C. (1946). *Measuring Business Cycles*. Cambridge, MA: National Bureau of Economic Research.

Parallels and Function

Introduction

This chapter is concerned with exploring the parallels in terminology and functional meaning between words used in ecology and economics. As Ecosystem Services represents a coming together of these two disciplines, it is important to be aware of the fact that similar terms employed by ecologists and economists may have entirely different functional meanings. In contrast, apparently different terms used by each group may be presenting a similar functional message. Here we will go through some broad concepts familiar to both ecologists and economists and try to identify potential sources of misunderstanding when specialists from each discipline begin to communicate the concepts inherent in Ecosystem Services with each other.

Methodologies

It is true to say that there are a suite of methodological procedures which are peculiar to the circumstances of ecology, involving the description, measurement and analysis of both the (ecological)

© The Author(s) 2019
S. Muddiman, *Ecosystem Services*,
Palgrave Studies in Natural Resource Management,
https://doi.org/10.1007/978-3-030-13819-6_3

population and the community [1]. In a similar vein, mainstream economics makes use of statistical analyses and observational studies to establish its theoretical framework [2]. But, in terms of the way in which the sciences of economics and ecology conduct their investigations, there are a number of differences and similarities.

Both economics and ecology make use of observations of existing systems to provide the raw data from which they aim to understand the underlying principles of the object of study. The observations made are used to develop a theoretical framework which has two aims. Firstly, to develop a narrative regarding the system being studied, and then to be able to predict future behaviour of the system when it is subject to different circumstances.

The observations made in both subjects relate to complex and active 'real life' systems. In subjects such as ecology and economics it is generally not considered feasible to isolate parts of the system in order to look at individual components and then reconstruct the whole. Because of the need to look at multiple factors at the same time, there is a high degree of specificity prevalent in the findings. The observations made are very much embedded within the context of the time and place that the study was undertaken. This means that such findings are often not generally applicable to a broader set of circumstances that the study itself.

For example, a chemical reaction will provide the same empirical results wherever in the world it is conducted. A study of the effect of grazing on a grassland will vary dramatically according to the time and place due to the unique combinations of season, climate, flora and fauna, etc. together with the details of the observational techniques employed. In order to try and reduce as much noise from the system as possible, ecological studies make use of controls in order to reveal underlying trends. However, the applicability of such trend to other circumstances cannot be assumed, without at least some additional information obtained from the alternative environment.

Economics is even more tied to the observational end of the scientific spectrum, as it is generally difficult, or in many cases impossible, to even apply control groups to a given phenomenon. It would not be a viable course of action to impose impoverishment on a part of society to investigate its reaction, or to enrich others in order to investigate the flow of wealth in a given situation. Any such attempts would result in such social upheaval that the effect of imposing the changes would completely obliterate the underlying tendencies which were trying to be discerned.

That having been said, mainstream economics has proposed significant 'economic engineering' based upon prevailing theories, although such modifications are not experimental in the true sense, i.e. testing the response to an action in order to prove or disprove a given hypothesis, but are considered to be management of the economy using an already established theoretical framework.

For example, in response to the 2008 financial crisis many central banks adopted a novel approach to addressing the resulting economic decline known as Quantitative Easing (QE). As described in chapter "Basics", most of the money in the economy is created by banks when they make loans, one result of the events of 2008 was that banks curtailed much of their lending activity (as low rates of interest meant it wasn't profitable for them to do so). This resulted in a decline in the amount of 'new' money being created.

This, together with the continued repayment of existing loans by borrowers meant that effectively the total amount of money in the economy was shrinking. In response to this phenomenon central banks in Europe, Japan the USA and UK adopted a scheme to create new money through QE.

The mechanism works by Central Banks purchasing financial assets, such as government and corporate bonds from the private sector. These bonds were paid for through the creation of new central bank reserves. Governments and the corporations approved to issue bonds would sell them to the Central Bank and in exchange, they would receive deposits (money) in an account at a retail bank. As a result of this transaction the bank would receive a new deposit (a liability from it to the seller of the bond), and a new asset—reserves at the Central Bank. Through the use of this mechanism, QE had the effect of both increasing money in Central Bank reserves and the deposits in commercial bank accounts.

The scheme, however, failed to provide a boost to GDP despite generating large amounts of cash. This was because the money created was used to make purchases from the financial markets, meaning the newly created money went directly back into the same financial markets, with the effect of boosting bond and stock markets nearly to their highest level in history. The Bank of England estimates that QE in the UK boosted bond and share prices by around 20%. According to mainstream theory, this should have made people feel wealthier, leading to an increase in spending, resulting in a knock-on growth in economic activity. However, as most of the stock market is controlled by a minority of already wealthy individuals, most families saw no immediate benefit from QE, whereas the richest 5% of households would have each had up to £128,000 of cash available directly as a result of the Bank of England's actions [3].

The Bank of England estimates that the first £375 billion of QE in the UK led to 1.5–2% growth in GDP. In other words, through QE it takes £375 billion of new money just to create £23–28 billion of extra spending in the real economy. This is clearly a very inefficient use of resources if the stimulation of spending is the aim of the exercise (Fig. 1).

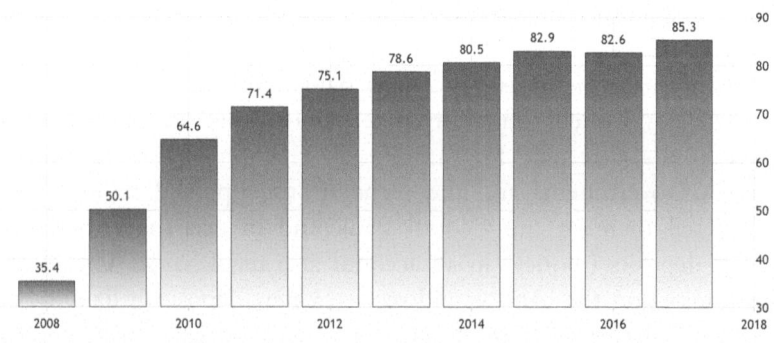

Fig. 1 UK debt to GDP (*Source* tradingeconomics.com | Office for Budget Responsibility, UK)

Another example of the application of theoretical knowledge having unintended consequences, drawn this time from ecology, can be seen in the case of the cane toad.

In the 1930s, Australia was suffering from the economic impact of a crop pest, the greyback cane beetle, affecting sugar cane. Cane toads were introduced from Central and South America as a predator in order to control the beetles.

However, the life histories and ecologies of the Cane toad and the beetle were not fully considered before its introduction, which led to a series of unintended consequences that proved disastrous.

Firstly, the beetle generally feeds high up on the cane, in areas which are generally inaccessible to the toads. Another key problem was that the beetles are diurnal, whereas the toads tend to be nocturnal. This means the two are rarely active at the same time, making predation of the beetles by the toad a rare occurrence.

In Australia, sugar cane is generally grown in areas with an arid climate. Such dry habitats are unfavourable to the toads, so they tended to migrate away from the fields into more suitable habitats. This phenomenon has proved to be the most significant problem as not only did the toad fail to control the target pest, but became a pest in its own right.

Cane toads are voracious predators, with a very broad diet, including small reptiles and mammals as well as invertebrate prey. They are also toxic to predators, and so make a poor choice of prey. These factors, together with a high reproductive rate have enabled them to spread widely and where they occur they have had significant adverse effects on local ecosystems.

In both ecology and economics the data obtained from observation are based upon a specific set of circumstances. The *modus operandi* of both disciplines is to attempt to apply a universal principle which fits the observed phenomena through the use of statistical analyses. Additional information can then either support, modify or refute the prevailing worldview. In both cases there are a series of plausible narratives which offer the option of multiple worldviews. The body of evidence will 'in aggregate' tend to support the narrative which most accurately reflects the actual phenomenon. There are few right or wrong answers in either discipline which can be applied universally, when approached from the basis of observational enquiry.

One element which is common to both disciplines is the way in which the prevailing narrative 'theory' is applied to the management of the topic in question. Both disciplines consider that there is an optimal state either of a particular aspect of the economy (interest rates, inflation level) or of an ecological system (species population, habitat or ecosystem) and that this state can be maintained through interventions which are determined by the relevant model.

An example from economics is the establishment of an optimum level of inflation (which is generally accepted to be at around 2%). Such inflation targets have been set by governments around the world. It is the responsibility of central banks to control inflation through the use of interest rates to either suppress or encourage spending. This, according to mainstream theory is a means of preventing the 'boom and bust' cycles observed in the economy. However, it can sometimes ignore other aspects of the economy which are often of greater immediate concern to its citizens, such as unemployment levels.

There is a current trend in conservation ecology to assign certain characteristics to habitats which determine them to be in 'favourable condition' on the basis of a range of indicators. Within the European Union the concept of favourable conservation status is central to the EC Habitats Directive [4]. Article 2 of the Directive, states that:

> Measures taken pursuant to this Directive shall be designed to maintain or restore, at favourable conservation status, natural habitats and species of wild fauna and flora of Community interest.

In reality, this means that it requires that the range and areas of a selected list of habitats, and the range and population of a selected list of species, should be at least maintained at their status when the Directive came into force or, where the status at that time was not viable in the long term, to be restored to a viable state.

Of course, the circularity in this requirement is obvious: any species or habitat which does not have a known status as of the implementation of the Directive cannot be listed. It requires an extensive body of pre-existing information in order to establish the precise criteria for favourable conservation status. It could be considered that such legislation has a counterproductive effect upon the ecological knowledge base within the European Union. All effort and funding is concentrated upon species which are listed within the Directive. Habitats and species which cannot be shoehorned into the necessary criteria are neglected. So, although the aims of the Directive may be met, it raises a question over the validity of these aims in the first place.

Recent studies of the concept of favourable conservation status in the European Union on a well studies group (birds) have shown that there is little consistency in the ways that such criteria are established and adopted [5]. If this is the case for such a well studied and popular group of animals then how can such a mechanistic approach be considered viable for the ecosystem as a whole (considering that much of the biodiversity of such systems is currently unrecognised)?

Certain phenomena have been observed in both economics and ecology which both disciplines consider to be consistently useful in producing theoretical models. These principles include the concepts of cycles, control mechanisms, hierarchies and energetics. By gaining an appreciation of the differences and similarities between the understanding of ecologists and economists of these phenomena, it is hoped that a common language will be possible, enabling the development of joint progress in a combined effort to develop a methodology which will reveal the underlying nature of the economics of ecosystems. Currently there is frequently a gulf in meaning of the same terms when used by the two disciplines, which can lead to failures in communication.

Cycles

Both Economics and Ecology have successfully identified features of their respective disciplines which fluctuate in a more or less regular way over time. Such fluctuations are generally termed cyclical in nature, that is, the point or state of the original observation will be repeatedly re-visited.

However, it is not straightforward to draw parallels or to make comparisons between cyclical phenomena even within either discipline, let alone to use them as a possible means of transferring concepts between the subjects.

In the first instance it is necessary to develop a more detailed appreciation that there are very different and distinct forms of cycles. Some are composed of discrete stages, which are quantitatively distinct and elements can be described independently of each other, with one stage leading consistently to the next. The most obvious example of this from the field of ecology is the life cycle of an insect with complete metamorphosis, composed of egg, larva, pupa and adult stages occurring in sequence, the adult of which produces eggs and so on.

This terminology has been employed in the economic sphere, particularly in relation to the development of consumer goods and the growth, maturity and decline of businesses and industries. The industry life cycle describes the way in which an industry or product develops over time through a series of distinct stages, introduction, growth, maturity, and decline.

Both of these forms of 'life cycle' are valuable descriptive tools, with clearly defined stages which can be followed in a logical manner.

Although both of these processes are termed life cycles, there is one obvious and significant distinction which must be borne in mind when considering any analogy between the two. Ecological life cycles are an element of a greater functional purpose, that being the reproduction and persistence of the species over time. In contrast, economic life cycles rarely lead to the persistence of the same Corporation (with some notable exceptions amongst the big brand names, although it is arguable whether the underlying ownership has persisted), rather it is the nature of economic cycles that they are driven more by competition from external sources, and internal evolution than a sustainable and ongoing use of resources over time.

In many respects, these life cycles are only an adaptive response to prevailing conditions which lie outside of the sphere of influence of the individual species or industries under consideration. In the case of a species, it is the natural variation in climate and seasons, etc. which stimulate the stage changes. With a business it is factors within the economic 'climate' which can affect the ability to invest or sell products. Unfortunately, the influences of these overarching variations are neither precise nor predictable as the cycles considered above. An unseasonable cold spell or an unexpected economic recession can disrupt these biological and business cycles.

When it comes to less well-defined cycles, there is an ability to identify a certain degree of periodicity within a series of observations, but the precision of such changes, and the precise mechanisms for their occurrence become more obscure.

In ecology, the observed variation in the population numbers (Population Cycle) of a species can be cyclical in nature, with times of abundance of any given species (either a plague or a 'good year', depending on the utility of the species concerned) and times of scarcity.

The underlying causes of these variations in population numbers are known in broad terms, and include factors such as resource availability, predation and parasitism levels, disease prevalence and the effects of climate. However, as a recent review of the subject notes 'Population cycles are one of nature's great mysteries' [6]. There are no universally applicable models and the mechanisms which ultimately cause cyclic variation in population numbers. This phenomenon must, therefore, be added to the catalogue of unknown components when considering biodiversity.

It is clear that ecological population cycles (which are based upon complex and unknown mechanisms) cannot be reliably manipulated in the real world. Experimental manipulation of such cycles is undertaken only in order to study the response of organisms, for the purpose of a greater understanding of the processes of response involved. The idea that climate manipulation should be carried out on a large scale for conservation purposes has been proposed in some circumstances as a response to apparent anthropogenic climate changes, but such proposals are few and far between. In contrast to this, economic cycles, based upon changes in the behaviour of individuals and businesses are routinely 'managed'.

The economy experiences a periodic fluctuation between periods of expansion (growth) and contraction (recession), the economic cycle. The current position within this cycle is generally measured by the rate of growth of GDP, interest rates, levels of employment and consumer spending.

The economic cycle is typically broken down into four stages: expansion, peak, contraction and trough (Fig. 2), but it is important to note that there is no fixed periodicity within the cycle.

Measured by changes in GDP, the average length of US economic cycles from the 1950s to present day is approximately 5 years. However, there is wide variation in the length of cycles, ranging from 18 months (1981–1982) to 10 years (1991–2001), according to the data collected by the US National Bureau of Economic Research (Fig. 3). In a similar manner to the population cycle of ecological studies, the cause of the economic cycle remains a cause of much debate.

Mainstream economics views the stages of the business cycle as being caused by a combination of the forces of supply and demand, the availability of capital, and expectations about the future.

The expansion phase of the cycle is characterised by confidence in the future, because there is anticipation of increases in pay and investment returns there is a tendency to make purchases. As the demand increases, businesses expand and hire new workers. The increase in consumer income further stimulates demand. The presence of inflation, which is controlled by Central Bank interest rates also stimulates demand by encouraging consumers to purchase immediately, before prices go up.

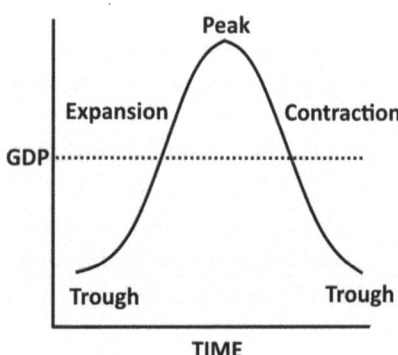

Fig. 2   The economic cycle

Change in GDP

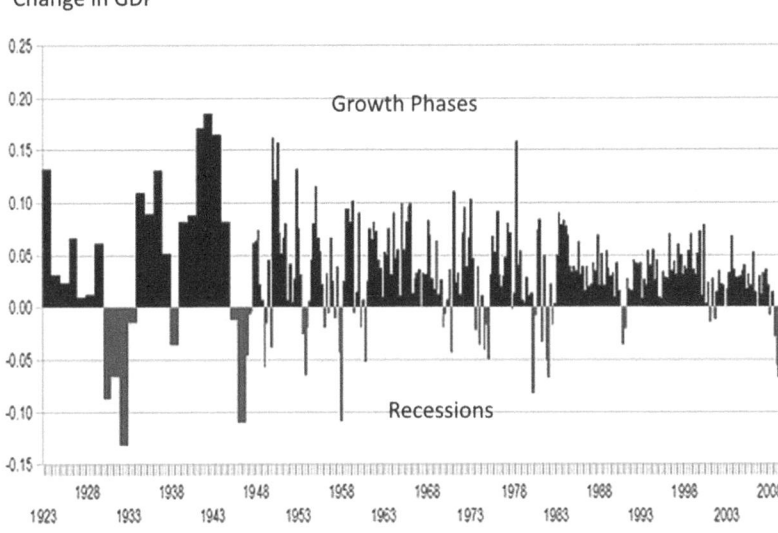

Fig. 3 US annualised changes in GDP (*Source* National Bureau of Economic Research)

The peak of the economic cycle occurs when there is an excess of money which outstrips the supply of goods and services, this leads to cause price inflation and overpricing of assets which are in demand, leading to an asset bubble. In response to increasing inflation, Central Banks raise interest rates. This also encourages savings and reduces the urge to obtain high returns through speculative investments.

The contraction phase of the cycle is characterised by a recession. Typically, this is triggered by high interest rates and inflation. At this point investment in industry is withdrawn, consumer spending declines and unemployment increases. The Central Bank responds by lowering interest rates and offering Government Bonds as a 'safe haven' against falling stock market prices. Simultaneously Governments borrow money, lower taxes and attempt to stimulate economic activity and employment through public sector works.

At the bottom of the cycle, consumer confidence is low. Central Banks attempt to improve confidence by supplying money through measures to encourage lending, and direct QE measures. Governments tend to reduce restrictions on Businesses though lowering corporate taxes, to encourage investment and industrial activity.

An alternative explanation of the mechanisms underlying the Economic Cycle is presented by the Austrian School of economics. In this view, low interest rates and the easy supply of money result in an imbalance between savings and investment. The expansion and peak phases of the cycle are characterised by low levels of saving associated with investment activity primarily focussed upon seeking a yield higher than the prevailing interest rate, rather than in sound business (malinvestment). Clearly, such activity is limited by the amount of money available to invest, and if more money is not forthcoming (which it is not, once malinvestment does not yield the expected returns) a recession then occurs. This leads to significant losses of money, which disappears from the system. Eventually resources are allocated only in profitable investments and savings, allowing a growth cycle to again occur.

The primary cause of this cycle is the increase in malinvestment until it becomes the overwhelming driver in the economy. Austrian economics assigns the initial expansion to the practice of fractional reserve banking, which encourages excessive lending and borrowing in a low interest rate environment. It is believed that the longer this excess production of money and malinvestment continues, the more extreme will be the following recession.

The recession is, in Austrian economic terms, a natural and inevitable process in which the economy adjusts from the earlier investment errors of the expansion phase and establishes an economy based upon sound investment. Continually expanding bank credit and low interest rates can extend the expansion phase of the cycle. But when no further investments can be found which provide adequate returns for speculative borrowers at the prevailing interest rates the contraction phase inevitably commences.

In addition to being critical of the Central Bank approach to the Economic Cycle, Austrian economists believe that attempts by governments to support asset prices, bail out insolvent banks and stimulate the economy with deficit spending enhance the level of malinvestment and serve to increase both the depth and duration of subsequent recessions.

The key difference between ecologists and mainstream economists when discussing any form of cyclical phenomenon is that the idea of a cycle in economic terms represents something which first and foremost is capable of management to reach a steady, optimal state. The majority of ecologists have expectations which are far more accepting of variation which cannot be avoided, and in many cases is actually a requirement of continued ecological health over the long term. Mainstream economists envisage sustainability as continuous and uninterrupted growth, without the variation seen in the economic cycle (famously paraded as 'no more boom and bust' economics), whilst Austrian economists see the mainstream approach to economic management as one which sows the seeds of its own cyclical demise. On the other hand, ecologists recognise that cycles themselves are the core of ongoing sustainability, that they are in fact a necessity.

Control

This brings us to how ecologists and economists view the concepts of control and management which involves two distinct perspectives. Firstly, there are the inherent controls present within ecological and economic systems, and secondly those interventions which are undertaken by mankind to influence the system being studied in order to deliver desirable outcomes.

In ecology, the type of habitats present in any location is determined by relatively few physical factors (see chapter "Basics"). However, the details of the component species and their interactions represent a snapshot in time of an ongoing series of relationships which drive not only the complex ecosystems we witness today, but are also responsible for previous and future adaptive changes in species morphology, physiology and life strategies.

When looking at habitats in these terms it becomes clear that simple overarching control mechanisms are rarely part of the ongoing evolutionary 'arms race'. Each species responds and reacts to threats and opportunities, with either positive or negative outcomes. Many of the supposed control mechanisms which we observe (such as the dynamic relations between predators and prey, or parasite and host) are in fact localised feedback mechanisms which represent only a tiny isolated portion of the whole complex.

This view is reflected in the Austrian Economic concept of Praxeology, where individuals act in accordance with their own economic needs and priorities, rather than being subject to an overarching set of rules. It is the decisions of individuals which appear to give the impression of some kind of mechanism operating when viewed in aggregate. This is similar to the patterns seen in large shoals of fish or flocks of starlings, which appear to be under the control of a single coordinating mechanism, whereas it is in fact the decisions and behaviour of individuals working in a way which maximise the chances of individual survival. The patterns and apparent coordination is very much from the individuals involved, rather than the imposition of rules from an external source.

Mainstream economics, however, views the economy as a system in which individual behaviour is less significant as a determining factor than a few simple parameters which can be manipulated by relevant authorities (be they Central Banks or Governments). These elements (interest rates, taxation levels, quantity of money, regulatory regime, etc.) are seen as the key levers which determine the economic outcomes and prosperity of the citizens subject to the management activities, carried out in accordance with mainstream economic theory.

The idea of control and management within ecological systems can superficially appear to have the same raison d'etre as the control theories of mainstream economics. In that the adjustment of relatively few parameters is undertaken in order to provide desirable outcomes, in this case usually associated with the conservation management of particular habitats or species. However, there is an important distinction which may not be immediately apparent to the casual observer. Conservation management originates from a series of very distinct aims and objectives which relate to a defined subset of the whole. Such action is generally associated with a habitat or species which is no longer thought to be viable in the long term without the proposed interventions. Conservation management is most definitely seen as an action of last resort, and most ecological systems are expected to operate without such interventions.

There has though, is recent times and in certain areas, been a certain degree of 'mission creep' with regard to human intervention into natural systems. Activities related to meeting arbitrarily generated conservation targets and the promotion of 'rewilding' (the reintroduction and translocation of species into their historical ranges) appears to accord more closely with the management concepts of mainstream economics, where a group of selected experts have their hands on the control levers of some form of mechanism (be it economic or ecological) and manipulate the whole to provide a perceived idealised outcome.

The concepts of control and management infer at least two components in a system, that which controls and that which is controlled. This opens up ideas of hierarchies within ecology and economics, another area which is worthy of some consideration.

Hierarchy

The recognition of hierarchies is something which has a fundamental place in biological sciences. Taxonomists place species in a hierarchical arrangement of related organisms according to Genus, Family, Order, etc. Ecologists study nested hierarchies, in terms of increasing ecological complexity (individual, population, community, habitat, ecosystem, etc.) or in functional terms (predators, producers, parasites, etc.).

At the level of individuals, ecology looks at the way in which an organism interacts with its surroundings. It identifies the adaptations and strategies that an individual uses for survival. This perspective places the individual at the centre of focus and views everything else as an influence upon it.

The next stage of investigation involves studies of populations of a species within a given geographical area, how it responds to both the environment and interactions with other species. The interactions between populations of different species within a given location, when they are taking place over multiple generations, are termed community ecology. Typical topics of study in community ecology include predator–prey relationships, parasitism and competition between species.

Habitat ecology looks at the relationships between multiple species and prevailing environmental conditions, forming a more or less stable complex of species.

This ecological hierarchy is nested in nature. Habitats are composed of communities, which comprise interacting populations, each population being composed of individuals. The other key hierarchical stratification used by ecologists is not nested in nature, this is the classification according to trophic levels, that is the position that each organism occupies in a food chain.

The start of most food chains are the primary producers, which convert non-biological materials into complex organic molecules, the most ubiquitous of such producers are members of the plant kingdom, which use solar energy to undertake the requisite chemical changes. The next level of the hierarchy are organisms which consume the producers, most commonly herbivorous species (primary consumers). These in turn are consumed by carnivorous predators (secondary consumers). In addition to this straightforward hierarchy are a range of side branches involving decomposers (such as many species of fungi), parasites and a range of omnivorous species, which alter their trophic level either on an opportunistic basis, seasonal availability of resources or scarcity of the preferred food supply. Such complexities typically result in a web of interactions (a food web) rather than a simple chain of consumption.

Recently a number of supplementary hierarchical systems have been adopted by ecologists. These all include the concept of presenting a species or species group as having a particularly disproportionate importance within the overall ecological system being considered. This can be seen as a natural human response to the unknown. By emphasising the value, significance and importance of what we do know, we are downplaying the significance of that we don't. This trend includes the development of the concepts of flagship, keystone and indicator species.

The concept of keystone species derives from studies in marine ecology [7, 8]. Keystone species are considered to disproportionately affect the stability of the habitat or ecosystem in which they occur. It is believed that the absence of such species would result in an exaggerated change in the composition and dynamics of its environment. This idea has become popular in conservation ecology as it allows a strong argument to be made for the conservation of particular species which are considered to be important. This allows decision makers and funding authorities to believe that the conservation of a known species is, by default, enabling the maintenance of a larger corpus of biodiversity.

The main criticism of the widespread use of the rationale of keystone species is that the actual veracity of such claims is rarely conclusively proven, it carries a certain political cachet without necessarily being backed up by ecologically determined fact. Furthermore, its breadth of application has meant that a precise definition of what constitutes an organism having the status of a keystone species does not exist.

The concept of a flagship species is one that is even further removed from ecological fact. It extends the idea that the protection of one species can benefit the greater good of a habitat or ecosystem into the realm of marketing, socio-economics and policy. Most species selected as flagship species fall into the category of 'charismatic megafauna', such as rhinoceros, tigers and elephants.

The key pitfall in the promotion of flagship species is that it runs the risk of misallocating funds at the expense of less charismatic, but more threatened, species (not to mention the 90% of biodiversity which remains unrecognised). Furthermore, if conservation efforts are not successful then local stakeholders who rely on the conservation efforts of flagship species for their livelihood may find themselves economically disadvantaged, even though there are significant needs in terms of the maintenance and conservation of other aspects of biodiversity within the same area.

Recent studies [9] have shown that even high profile flagship species tend not to be lifted out of endangered status through the publicity and funding received. The research suggests that the flooding of media and conservation literature with relatively few species only tends to convince the public that these species are not as rare as scientific studies suggest. The reasoning being if they can be photographed, filmed and researched so extensively then they must be relatively abundant. This, of course is likely to have a grain of truth in it, as we remain blissfully unaware of the true conservation status of the majority of biodiversity on the planet.

According to the United Nations Environment Programme, an Indicator Species is one 'whose status provides information on the overall condition of the ecosystem and of other species in that ecosystem. They reflect the quality and changes in environmental conditions as well as aspects of community composition'. This definition is very broad, and tends to conflate several distinct roles, some of which may be considered to be lacking in veracity.

If the sensitivities of an organism or group of organisms have been thoroughly researched then changes in such species abundance or behaviour may be of value in extrapolating to corresponding changes in physical conditions within the environment. This use of indicator species (or groups) is well established in aquatic regimes, where the presence or absence of certain invertebrate families can provide a broad indication of the degree of organic pollution present in a watercourse. As another example, the use of certain plants with specific habitat requirements and limited dispersal abilities within woodlands can provide an indication of the age of the woodland.

However, the selection and application of the concept of indicator species into less well researched areas blurs the lines between a species being a biological indicator in the true sense and it being promoted as either a keystone or flagship species, as previously discussed.

In the field of economics, hierarchies are often established in relation to the behaviour and needs of consumers. Although not strictly a purely economic concept, the idea of a hierarchy of needs, as epitomised by that of Maslow [10], represents a model which, consciously or not, drives both the economic decisions of individuals, but also sets priorities for policy.

This system is usually represented in the form of a pyramidal hierarchy of human needs and desires (Fig. 4).

The general principle is that the needs at the base of the pyramid represent priorities, until they are met, at which point the needs at the next

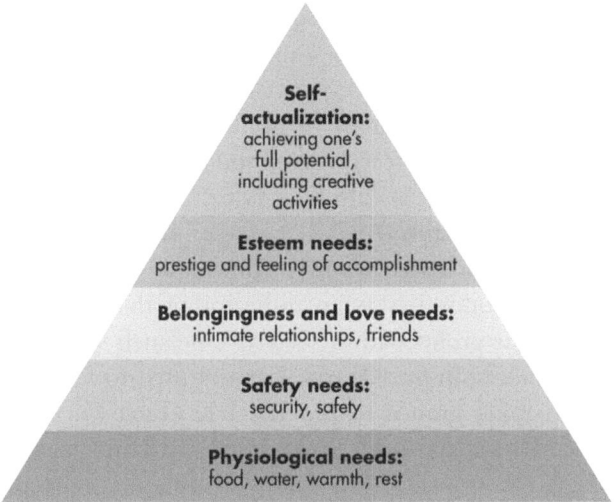

Fig. 4   Hierarchy of needs (Maslow [10])

level become the most pressing. When conceived, this hierarchy was initially seen as a more or less linear progression of changing priorities, however this was later modified, in that the satisfaction of a need is not an 'all-or-none' phenomenon, so that a need does not have to be completely satisfied in advance of the next.

At the most basic level, physiological needs such as food, drink, shelter, clothing and sleep represent the fundamental priorities. The next series of needs are those associated with safety, such as protection from elements, security, order, law, stability, freedom from fear. After physiological and safety needs have been fulfilled, the third level of human needs is social. The need for interpersonal relationships and being part of a group motivates behaviour. Beyond this stage are self esteem, and the desire for reputation and status amongst one's peers. The final level in the hierarchy is the realisation of personal potential, described as 'to become everything one is capable of becoming'.

These motivations are used as aspirational themes both in terms of the marketing and advertising of products, representing a means of influencing the spending behaviour of consumers, but also in respect of how governments use tax revenues and how they determine priorities for public spending.

In some respects this hierarchy of needs is reflected within the classification of Ecosystem Services, where Producing services tend to fulfill physiological needs, whereas cultural services are more focussed on the upper levels of the pyramid.

Social stratification, based upon income and occupational activity is often used by economists as an indicator of the current economic state of a particular group within society. One such social hierarchy is the National Readership Survey social grades used to identify different income demographic groups within the UK (Table 1). An additional use of this hierarchical approach is that it can track and assess the effects of policy, such as the so-called 'trickle down' economic concept of mainstream theory. In this idea, an injection of wealth at the top of the hierarchy stimulates the economy, encouraging employment and spending, thus raising the standard of living of those at lower income levels.

Table 1 Social grades

Grade	Social class	Chief income earner's occupation	Frequency in 2008 (%)
A	Upper middle class	Higher managerial, administrative or professional	4
B	Middle class	Intermediate managerial, administrative or professional	23
C1	Lower middle class	Supervisory or clerical and junior managerial, administrative or professional	29
C2	Skilled working class	Skilled manual workers	21
D	Working class	Semi-skilled and unskilled manual workers	15
E	Non working	Casual or lowest grade workers, pensioners, and others who depend on the welfare state for their income	

UK social grades by chief income earners occupation
Source NRS (2008)

Economic stratification and the formation and coalescence of a hierarchy over time is aided by a phenomenon known as the Cantillon effect [11]. This identifies the effect of changes in the supply of money on prices. Such changes are not uniform throughout the economic hierarchy because the change in money supply has a specific point where it is introduced and follows a specific path through the economy. The first recipient of the new supply of money is able to make expenditures in advance of any increase in prices, so they benefit disproportionately. However, later recipients of the money (as it moves through the economy) see less benefit from as the effect of the increase in money supply has already lead to higher prices.

Some of the hierarchies established in both economics and ecology represent the movement of energy through each system, in modern technological economies, the energy involved in driving industrial growth is of paramount importance, in ecology it is energy which is transferred within the food chain/web. Therefore, the flow of energy can be seen as a concept which is common to both disciplines.

Energetics

The concept of energy returned on energy invested (EROEI) or energy return on investment (EROI), is common to both ecology and economics when considering the energetics of any given system. It is the ratio of the amount of usable energy which is gained from an energy providing resource (either the sun, or food in ecology and the various forms of energy generating resources: solar, wind, nuclear, oil, etc. in economics) against the amount of energy used to obtain the resource and exploit its inherent energetic value. When the ratio is less than or equal to one, the amount of energy being expended in finding and using the energy is actually less than the benefit being received. In ecological terms this means that the organism involved will effectively starve to death, in economics, it makes the exploitation of that particular resource non-viable. In economics, a fuel or energy with an EROEI ratio of at least 3:1 is considered to be viable as an energy resource, in order to allow for subsequent inefficiencies in its use, and to allow profits to be made.

In ecological systems, the flow of energy through a food chain or web usually begins with the capturing of the energy of the sun through the photosynthetic activity of green plants (producers). The energy which is accumulated by these organisms is harvested by herbivores or omnivores feeding on plant material (primary consumers). Some of the energy obtained us used to build up biomass, which can, in turn, be harvested by carnivorous species (secondary consumers).

This description is essentially the hierarchical food web described in energetic terms. At each trophic level, only about 10% of the energy obtained is converted into material available for the next level in the chain. Therefore, primary consumers get about 10% of the energy gathered by producers, whilst secondary consumers get 1% of this original energetic input. Although this serves to place a limit on the number of links within any given food chain (as the amount of biomass available for consumption is decreasing), it does not mean that carnivores are less energetically viable than herbivorous species. The key is the equation which relates the amount of energy expended for a given amount received. This balance operates universally across all trophic levels, from plants being unable to obtain sufficient sunlight, to a wolf unable to obtain prey.

Since the industrial revolution coal, oil and other geologically modified plant materials have been the foundation of most economic activity. Rather like ecological consumers, industrial resource companies are seeking to gain an energetic benefit from previous production which is now sequestered in geological formations. This involves an expenditure, which is seen in terms of financial investments in machinery, transportation infrastructure, refining and storage facilities, etc. However, when looking at the energetic equation, these costs can be better evaluated in terms of the energy expended in their creation. The energy expended in terms of machinery production, movement, etc. has to outweigh the energetic value of the resource obtained. Unfortunately, mainstream economics establishes a surrogate measure for these factors via the provision of a financial measure of value. This measure does not necessarily provide a straightforward answer whether a resource is worth extraction for two reasons.

Firstly, purely fiat money is elastic in its value, dependent upon how much is in circulation, so changes in the quantity of money can have unforseen effects upon the monetary equation in relation to energy costs and benefits. This is a totally different scenario to the cost/benefit analyses within an ecological system. Here, the benefits are stable and fixed in nature, even when some of the benefits are not realised until a future date (through storage of food items, or storage of resources through body fat).

The second issue with the use of financial measures is that it allows for the availability of credit. This can distort the equation towards extraction of non-viable resources (with associated environmental effects). Future default on repayment of the credit received does not result in the full realisation of the consequences of the actions (energetic and environmental), as only a financial loss is recognised within the economic system.

At the beginning of the twentieth Century, oil was plentiful and easily accessible. Oil extraction could be carried out with a relatively low amount of input (e.g. high EROI). This gave petroleum products a great advantage as a source of energy, there was no need to search for alternatives and led to a surge in growth of the petroleum industry as a whole, together with other industries which made use of the store of energy which oil provided.

However, as the most easily accessible sources of oil began to be used up, increasing energy was required in order to obtain the same quantity as before. This resulted in a decline in EROI, which has continued to the present day [12] (Fig. 5).

Supply and demand economics would suggest that as oil and its derivatives became more expensive to extract, the price of these commodities should rise in order to cover the increasing costs. However, the price of oil has not been able to rise to the extent required. This is because oil has not been used with the required degree of efficiency to allow the productivity of oil using industries (and other consumers) to rise sufficiently to allow higher prices to be paid. As consumers, humanity is burning off its stores of energy without using it in a way which

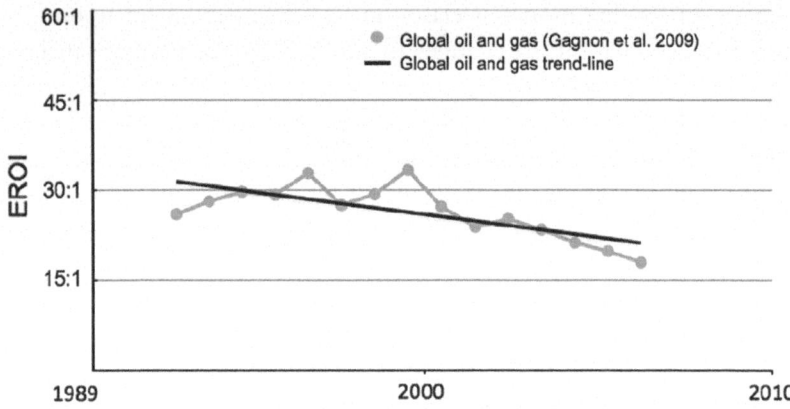

Fig. 5 Global oil and gas EROI (*Source* Gagnon et al. [12])

creates wealth. In the light of disproportionately low prices, the oil industry is increasing investment in exploration and production by borrowing increasingly large sums of money. This borrowing will at some point either need to be repaid, or defaulted on. The financial mechanisms which allow such a situation to persist do not operate within the ecological sphere, it is not possible for an organism to run an energy deficit in this manner.

Mainstream economic theory is intimately associated with the circumstances surrounding the presence of cheap, abundant energy resources. It should be noted that historically this situation is rarely encountered and, as noted above, it is a situation which is not sustainable due to the increasing burden of debt which it engenders. Therefore, it is worth considering to what degree mainstream economic theory has been skewed by the prevailing energetic regime under which it has been developed.

The mainstream economic policy approach of increasing public works expenditure to offset recessionary economic trends fails to operate without cheap, abundant energy supplies. The aim of mainstream economics is to continue an upward trajectory of growth (usually measured as GDP). In order to achieve this, it further assumes that future energy supplies will continue to be cheaply available in the required quantity to meet growth in demand.

As we have seen above, energy sourcing is becoming increasingly energy intensive, with the shortfall being met by increased debt, using money as a surrogate for energy. The belief is that this energetic debt will be repaid through enhanced efficiency in the use of existing resources. However, the data does not currently support this view, as productivity is, in fact declining (Fig. 6).

These examples show clearly the difference between economic and ecological energetics. In an ecological system, energy deficits cannot persist for any length of time without the demise of the species involved. In mainstream economics, however, it is possible for industry to continue on the basis of a long-running energetic deficit which is hidden through the financial mechanisms of money creation and interest repayments.

This dichotomy between the fundamental operation of ecological systems and economic ones will be brought into clearer focus in the next Chapter, which deals with the valuation of ecological systems in the context of mainstream economic theory.

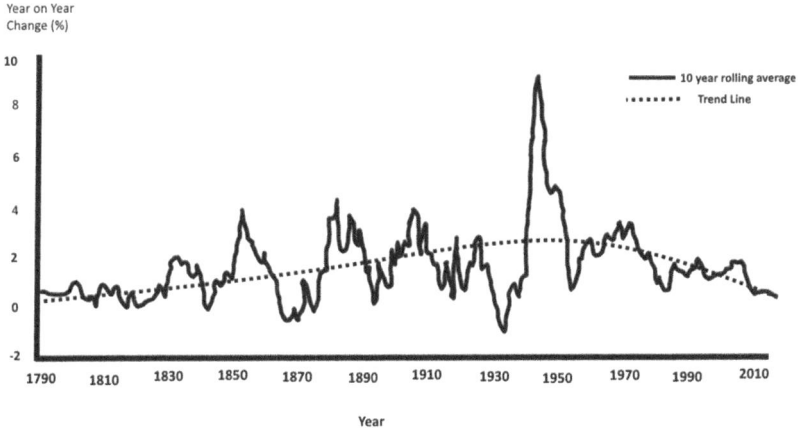

Fig. 6   Productivity GDP per capita (*Source* Bawerk.net)

References

1. Southwood, T. R. E. (1978). *Ecological Methods*. London: Chapman & Hall.
2. Samuelson, P., & Nordhaus, W. (1948). *Economics*. New York: Irwin/ McGraw-Hill.
3. New Economics Foundation. (2013). *Strategic Quantitative Easing*. https://neweconomics.org/2013/07/strategic-quantitative-easing.
4. European Union. (1992). *Council Directive 92/43/EEC on the Conservation of Natural Habitats and of Wild Fauna and Flora*.
5. McConville, A. J., & Tucker, G. (2015). *Review of Favourable Conservation Status and Birds Directive Article 2 Interpretation Within the European Union* (Natural England Commissioned Report NECR176).
6. Myers, J. H. (2018). Population Cycles: Generalities, Exceptions and Remaining Mysteries. *Proceedings of the Royal Society B: Biological Sciences, 285*(1875), 20172841.
7. Paine, R. T. (1969). A Note on Trophic Complexity and Community Stability. *The American Naturalist, 103*(929), 91–93.
8. Scott Mills, L., Soule, M. E., & Doake, D. F. (1993). The Keystone-Species Concept in Ecology and Conservation. *BioScience, 43*(4), 219.
9. Courchamp, F., Jaric, I., Albert, C., Meinard, Y., Ripple, W. J., & Chapron, G. (2018). The Paradoxical Extinction of the Most Charismatic Animals. *PLoS Biology, 16* (4), e2003997.
10. Maslow, A. H. (1987). *Motivation and Personality* (3rd ed.). Delhi, India: Pearson Education.
11. Cantillon, R. (2010 [1755]). *An Essay on Economic Theory*. Auburn, AL: Ludwig Von Mises Institute.
12. Gagnon, N., Hall, C., & Brinker, L. (2009). A Preliminary Investigation of the Energy Return on Energy Investment for Global Oil and Gas Production. *Energies, 2*, 490–503.

Valuing Ecosystems

Introductory

The main driver for the development of Ecosystem Services as a systematic framework is to enable decision makers to formulate policy decisions based upon a sound understanding of the value which mankind derives from the natural world. It is an attempt to permit rational cost-benefit analysis of human activities which affect natural systems. For this framework to operate there needs to be a common measure when considering costs and benefits otherwise there is always the danger of comparing 'apples with oranges'. Establishing such a common baseline involves the process of valuation. In this chapter we will look at approaches to this valuation of the environment in general, and Ecosystem Services in particular.

Before beginning this task, it is important to realise that 'value' as a term holds two similar, but distinct meanings. Firstly, when something is valued it can mean that it is held in high regard or considered to be of importance, such a meaning not only relates to material goods and services, but also to morals and ethics, where an individual holds certain values. When Ecosystem Services are approached with this meaning of

© The Author(s) 2019
S. Muddiman, *Ecosystem Services,*
Palgrave Studies in Natural Resource Management,
https://doi.org/10.1007/978-3-030-13819-6_4

value in mind then there is an almost immediate consensus that they are of value, but the quantification of such value remains intangible.

The second meaning of value is in terms of monetary worth, this is essentially an economic exercise. This approach uses the processes of trade, or cost of substitution in order to assign a monetary value to an item or service. The practice of such valuation is far from straightforward as the whole concept of economic value presupposes that the subject of the valuation process can be pegged against some form of benchmark. When it comes to the valuation of Ecosystem Services the process becomes complicated by the fact that it is essentially a free service which has been granted through serendipitous circumstances. An Ecosystem Service was not manufactured or created with an end purpose in mind. Due to this, there are no labour or development costs to be incorporated into the valuation process, it is purely the cost of replacement or the value to consumers which will dictate the conclusions drawn.

Currencies alter in value over time, as a result of inflation, speculative activity and their value relative to other currencies. The variation in the comparative value of the British Pound Sterling and the US Dollar is used as an example of these fluctuations (Fig. 1). There is no stable currency on the planet, particularly under the prevailing fiat system, where the amount of each can be expanded at will. The question of the units to be used in the valuation process is therefore something which needs to be considered with care. If a fiat currency is used to assign a value, then the worth of a service may appear to be great at one moment in time, but over a period of months or years it could appear to become less and less significant if the currency it has been valued in becomes debased. The decline in purchasing power of the US Dollar over time is illustrated in Fig. 2.

Another question raised by attempting to determine what something is worth is: Who is the value assigned to? Monetary value is only what someone is willing to pay. In theory the highest bidder of a good or service values it more highly that other consumers. This approach to valuation is, however, completely inappropriate when considering biodiversity or Ecosystem Services. as those who are most dependent on a service are frequently those with the least ability to pay for it.

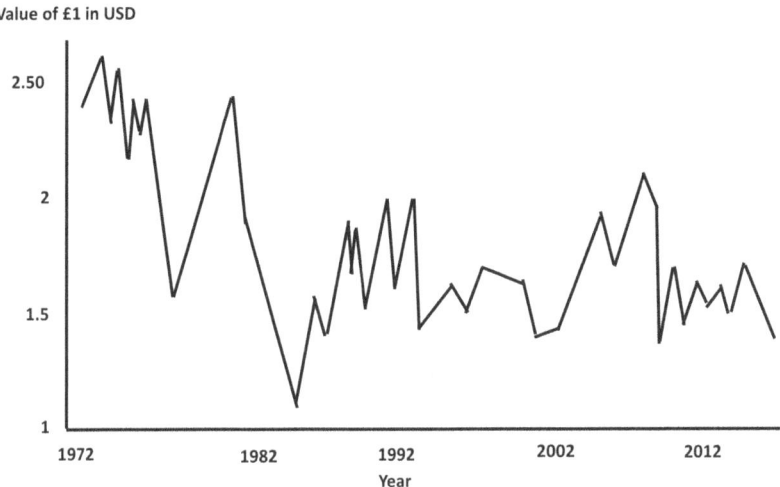

Fig. 1   Value of GBP vs. USD (*Source* Federal Reserve Bank of St. Louis)

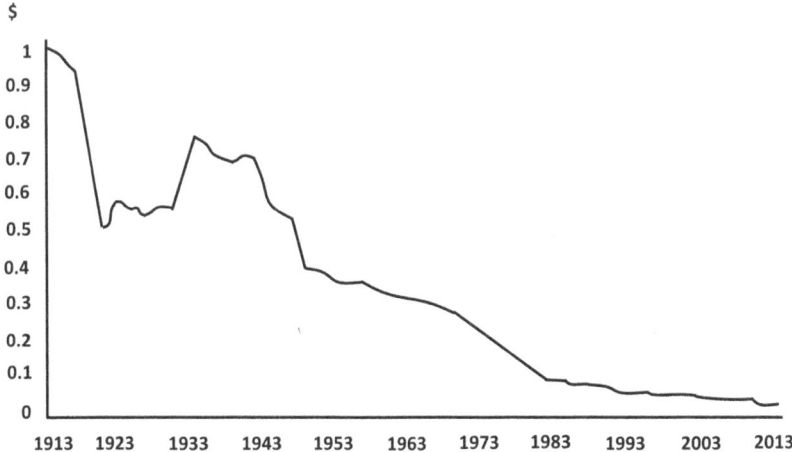

Fig. 2   US Dollar purchasing power change over time (*Source* US Bureau of Labour Statistics)

The stakeholders who regularly make use of a service are likely to be easily 'outbid' by a corporation or government. This is particularly true when considering that these institutions have access to almost unlimited credit lines within mainstream economic practice.

The most common case requiring an estimate of value is when goods or services are being bought and sold in a commercial setting. This transactional value is not something which can be applied in the case of Ecosystem Services, as the question of ownership of the services provided is not something which is being transferred. Valuations can also be applied in cases where there is a need to make comparisons between options. For example whether one is getting a 'fair deal' when undertaking exchanges of produce is most easily determined by assigning a value in a common currency to both items. This is closer to the rationale for valuing Ecosystem Services, in other words finding a common denominator by which the comparative value of an Ecosystem Service and an alternative may be measured. But it still fails to take into account the variation in value experienced by different stakeholders, and still allows for debt base financing to essentially overwhelm any other concern (Fig. 3).

We have already demonstrated that our knowledge of biodiversity is very limited. This is an important fact, as knowledge can affect value. If there is a mismatch in knowledge regarding an object or a service then there will be significant differences in valuation. More knowledge surrounding a subject will not only allow a more accurate valuation to be made, but generally a deeper understanding of the utility of a good or service will also enhance its value. For example, a bidder who knows the true rarity of, say, a watch, will know much more accurately what others would be prepared to pay for it. But conversely, if the watch is broken,

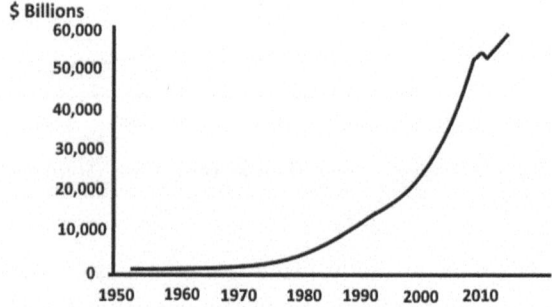

Fig. 3   US credit market debt (*Source* Board of Governors of the Federal Reserve)

and this fact is not known then it will be overvalued in the eyes of the purchaser. This factor is of critical importance in terms of Ecosystem Services. As we are unaware of much of the biodiversity that drives the Ecosystem Services we are concerned about, how can we possibly have knowledge of the actual rarity or functional status of that which we are attempting to put a price on?

Trends and fashion can also have a great bearing on how goods and services are valued. Even within taxonomic studies, there is a clear bias towards research on the more fashionably popular groups of plants and animals [1] with a disproportionate amount of effort (and resources) being applied to relatively few groups (Fig. 4).

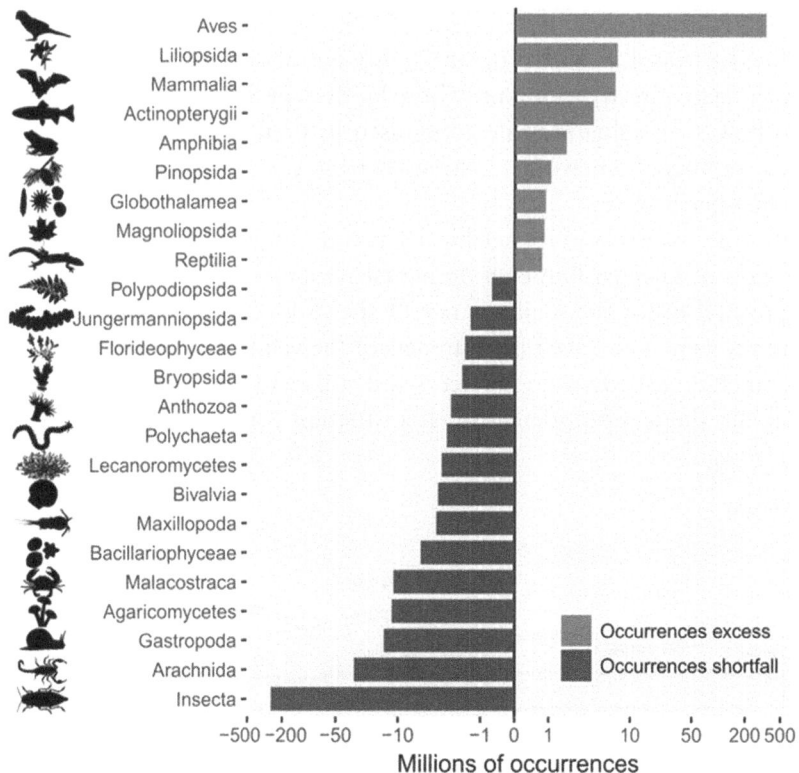

Fig. 4   Research effort in taxonomic groups (*Source* Troudet et al. [1])

Over the last decade or so, one of the greatest drivers in regard to environmental studies has been the issue of climate change. This has acted as a vehicle to publicise and attract funding for a wide range of projects and studies which could easily have been framed in other terms, but the use of climate change as a driver has led to an increase in the perceived value of the study. For example, works on agricultural adaptation to changing climates could just as easily be framed as studies of patterns of agriculture in existing climatic zones. Studies on flood relief and water resource management also tend to be framed in terms of global changes. It is the concerns surrounding climate change which give such studies a high profile and increases their perceived value. The amount spent on climate-related research, science, technology and aid has increased significantly. Figure 5 shows the amount spent by the United States on such projects since 1993.

Land ownership and property rights are another contentious issue which relate directly to not only how Ecosystem Services are valued, but also how such valuations are recognised in policy decisions under different political frameworks. Land ownership can be perceived from two distinct perspectives.

On one hand, land ownership by private individuals can be seen as a means of safeguarding environmental resources, as ownership confers responsibility for the maintenance of the property. This view, however requires owners to have full awareness of the value of the resources present (including Ecosystem Services) and sufficient capacity to be able to maintain them appropriately. Such a situation could lead landowners of

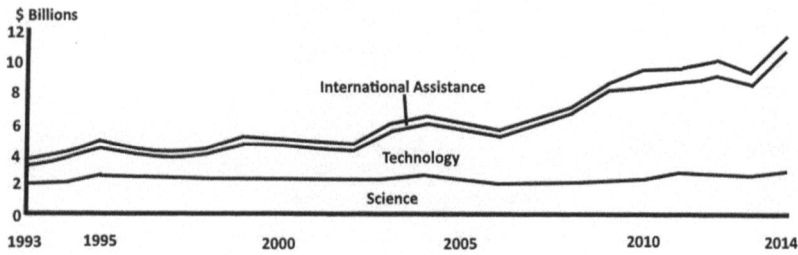

Fig. 5 US climate related research spending (*Source* US Office of Management and Budget and Congressional Research Service)

services which are of community importance to raise a levy (or receive government subsidies) to pay for works and to prevent economic pressures requiring Ecosystem Service damage or loss as a result of change in land use. For example, if a landowner needed to convert all available land into arable cultivation at the expense of a wetland which provided flood relief, due to the low price of crops.

The second approach involves a belief that the owner of property will be concerned primarily with maximising the profit from it, thereby making community significant Ecosystem Services vulnerable when placed into the hands of private ownership. This results in the appropriation of important areas by some form of government authority. In turn, this will result in the use of taxation to employ public sector employees for the maintenance and management of the land, under direction from policy decisions made remotely from the location. Of course, this raises an additional level of insecurity, as once an environment becomes dependent on public funding it also becomes vulnerable to changes in priorities at state level and the prevailing economic climate at a national level.

In addition to these perspectives, it is worth noting that quite often, even private land ownership is not secure from government compulsory purchase, when public works are proposed which are seen to be in 'the national interest'. Such works are often derived from the mainstream economic theory that economic downturns may be avoided through enhanced public funded activity.

Intrinsic Value

One point of view holds that the natural world has a value which is intrinsic and unquantifiable. This approach uses the ethical and moral definition of value as the basis of how we should regard the environment. It is not concerned with valuation in the economic sense of the term.

In many ways this approach to the environment is associated closely with the cultural aspects of Ecosystem Services. As an approach it does not lend itself to quantitative analysis, but it is a powerful driver in the formation of policy and political orientation with regard to environmental issues.

Much of the approach relating to intrinsic value places itself at odds with concepts of mainstream economics [2]. Although identifying the impossibility of placing a financial cost on things which have value in a moral or ethical sense, there is generally a failure to identify the fact that alternative economic models may result in different outcomes. For example, Austrian economics treats the actions of each individual within the economic sphere to be a result not only of a series of financial priorities (paying debts, purchasing essentials for life etc.) but it also considers individual economic decisions to include elements of the moral and ethical make up of the individual. Of course, to express such preferences, it is necessary to achieve levels of economic freedom which are not available under a more controlled economic system.

Within the area of environmental philosophy and ethics, intrinsic value is defined as the value of nature in and of itself, whilst the value of nature to humanity is described as instrumental value [3, 4]. Ecosystem Services, as a manifestation of instrumental value, are therefore placed in counterpoint to intrinsic value.

However, it is not possible to ascribe a value of any sort (monetary, ethical, moral or spiritual) without the presence of a conscious awareness of the element being valued, together with a sense of its importance. Therefore, even intrinsic value has to incorporate the human mind within its definition. Because of this the concept of intrinsic value appears to be mis-framed.

It would be more constructive to consider all environmental valuations to be instrumental in nature, as without the involvement of human consciousness they cannot be recognised or valued in any sense. This is the situation for the 90% of biodiversity which is currently outside the sphere of human knowledge. Despite this, it could be considered that the awareness of our ignorance in this field could actually serve as a vehicle for valuing the unknown.

The known environment can be considered as having either utility value, social value or spiritual value. Utility value can be considered in terms of our knowledge and use of the environment, including the majority of Ecosystem Services as currently framed. Environmental issues and features can also be made use of for social and community cohesion. In many cases an environmental cause can rally strong support (as discussed in relation to Flagship species), a physical feature of the environment can also bring a community together as a focal point in the landscape or as a significant part of a groups historical/psychological narrative. Whether such elements have true spiritual meaning or are acting as a proxy for social cohesion and organisational purposes is a point of contention in many cases. True spiritual value is based upon an individual's own relationship with the natural environment and as such has little power or influence in terms of either economic or social moves to either conserve or effect change at a larger scale. The love of a single individual for an area or feature, if not supported by popular opinion, status or influence, is given minimal consideration in valuing the environment at a wider level:

> To see a world in a grain of sand. And a heaven in a wild flower, Hold infinity in the palm of your hand. And eternity in an hour. William Blake

When intrinsic value is invoked as a consideration in policy, its adoption and promotion can result in a coercive approach to legislation which is driven by moral and ethical considerations over the needs of individual stakeholders. For example, in some cases restrictions may be placed on the ability of an individual to conduct certain activities which are of importance for their well-being, such as cutting timber or cultivating certain areas of land. These restrictions may be perceived as an imposition by those who hold moral superiority and may lead to community and social fragmentation rather than cohesion. In such cases the 'greater good' becomes nothing less than a tyranny. In order for intrinsic value to operate as a policy tool it needs to be clearly defined and have an agreement from all stakeholders for it to operate effectively and make a tangible difference.

Natural Capital

Natural Capital is a concept which has been developed as a parallel to the economic meaning of the term. In mainstream economics capital refers to financial assets together with the tangible factors of production including equipment used in the production of consumer goods.

However, the recognition of capital can vary according to the underlying economic theory. Mainstream economics recognises capital as including financial assets which are debt based, in other words borrowings can be recognised as capital rather than merely as debt. In contrast to this, the Austrian School of economics believe that capital is accumulated through the saving of a portion of the profit of productive labour. In this way the product of labour can either be consumed, traded or converted to capital.

Furthermore, the Austrian view is that capital is not a viable concept outside of the sphere of accounting [5]. In other words, capital items must be able to have a tangible value applied to them. One criticism of other economic systems by the Austrian school is that they often fail to include the natural environment in these calculations. Thus, the Austrian school introduced, in its definition of capital, the fundamental idea of natural capital in a truly economic sense.

When interpreted in this manner, this means that the value of the natural capital must be taken into account when considering not only environmental valuation practices, but also in the realm of purely economic cost-benefit analysis and when calculating the profit and loss of any economic operation. This element could be calculated as the sum of the value of all the services received, multiplied by the length of time the natural capital functions to provide the service. In some cases, natural ecological development of habitats or other factors mean that certain services are only available for a limited amount of time. For example some wetland habitats are transitory in nature and at certain stages may have both flood relief and water purification benefits, but natural habitat development may reduce or cease these services over a period of time.

The European Environment Agency (www.eea.europa.eu) defines natural capital as 'An extension of the economic notion of capital (manufactured means of production) to environmental "goods and services". It refers to a stock (e.g., a forest) which produces a flow of goods (e.g., new trees) and services (e.g., carbon sequestration, erosion control, habitat)'.

This concept is promoted to advocate for the conservation of ecosystems, framing their value in the language of mainstream economics. Ecosystem Services are considered to be the products of natural capital accumulation. By following the economic model these products of capital may be used, as long as there is no depletion of the capital itself. Unfortunately, the measurement of the components of natural capital, and whether the use of an Ecosystem Service results in their depletion is not something that can be carried out in a detailed manner, bearing in mind our limited knowledge and understanding of biodiversity and ecosystem function.

For example, the extraction of timber from a natural woodland ecosystem may appear to be 'sustainable'. But to all intents and purposes it is removing a resource and making it unavailable for other elements of the ecosystem. There will be less material for decomposers to act upon, thereby reducing free nutrients in the soil over the long term and also potentially reducing the diversity of the species involved in the decomposition process. Although superficially, such effects may appear marginal, there is no quantitative data which supports the idea that such activities are in fact sustainable over an extended period of time. Things may appear to be the same year over year, but the intricacies of such actions cannot be definitively mapped and understood on an extended timescale.

The European Environment Agency, in its explanation of natural capital moves significantly away from a purely economic perspective, including elements which cannot be valued or accounted for. Therefore, the term natural capital is a misnomer in this case. The EEA considers that there are four core stocks of capital: manufactured capital (e.g. machines and buildings), human capital (e.g. people, their skills and knowledge), social capital (e.g. trust, norms and institutions) and natural capital (e.g. minerals and Ecosystem Services). It then goes on to consider that capital in the economic sense is just a medium of exchange between these core stocks.

Not only does this approach invert the entire concept of capital, as developed by economists and economic theory, but it expands the definition to include the sources of all tangible materials, all naturally sourced resources, together with labour and the entire social structure. For example, mineral reserves are considered to be natural capital in the existing definition. This begs the question 'what Ecosystem Services do minerals provide mankind, whilst they lie in an unextracted and unrefined state under the ground?'. Of course, these mineral reserves are only of utility value when extracted, refined and made into useable products. The use of the resource is directly a part of the production process. Therefore, such resources are just that, resources, not capital items.

Furthermore, the organisation asserts that:

> Natural capital is the most fundamental of the forms of capital since it provides the basic conditions for human existence, delivering food, clean water and air, and essential resources.

This statement is only likely to be true in extremis, it is not supported by current evidence. Humanities population has been growing at the same time as a reduction in available natural capital. This gives a strong indication that either the whole concept of natural capital is an erroneous analogy with economic capital, that it is not natural capital which provides the conditions for human existence or the concept has been subject to an exaggerated interpretation in order to justify a particular policy agenda.

By linking natural capital with the valuation of Ecosystem Services, there is an implicit requirement for a degree of ignorance of the actual functioning of ecosystems in order to be capable of delivering desired policy outcomes. For example, if an ecosystem was sufficiently well known and understood and it was concluded that there were no tangible Ecosystem Services emerging from it, then this ecosystem would not constitute natural capital. Of course, this argument is avoided by extending the definitions of natural capital and ecosystems so broadly that it is extremely unlikely that any ecosystem (broad definition) would not provide any Ecosystem Service, therefore they all qualify as natural capital (broad definition). Although such an approach permits the achievement of certain policy objectives, it fails to provide an actionable quantitative valuation methodology and discourages gaining a deeper and more comprehensive understanding of ecosystems. Once policy objectives have been met then it is a matter of 'mission accomplished', rather than continuing to study a system to determine its true value. This is particularly significant in cases where there is a possibility that the value of systems based upon current knowledge have been under or overvalued to a functionally important degree.

Natural capital in the true sense of the term is very much concerned with the utility value of Ecosystem Service provision. At its root, natural capital represents an accounting tool which is currently inapplicable for several reasons:

- Current definitions are too wide and include intrinsic values which cannot be quantified.
- The definition encompasses elements of the environment which fall outside the economic definition of capital.
- There is insufficient current knowledge to quantify those elements which can validly be described as natural capital.
- The current fiat monetary system devalues capital (including natural capital) over time as a result of policy-induced inflation.

TEEB Approach

The issue of valuing Ecosystem Services following the principles of TEEB outlined in chapter "Basics" has been addressed in a report [6]. This document comprises a series of separately authored papers which cover the concepts and methodologies for a proposed economic valuation of biodiversity and Ecosystem Services. The document comprises six topic chapters and a seventh, which reiterates the key findings of the previous chapters and offers linkages to local and national policy.

The first chapter 'Integrating the ecological and economic dimensions in biodiversity and Ecosystem Service valuation' is primarily concerned with establishing a framework for undertaking assessments of Ecosystem Services, so that such exercises can be undertaken in a consistent manner. It establishes that an ecosystem assessment should:

- Clearly define the spatial extent of the assessment and the timescale in which it is valid.
- Define the ecological and physical composition of the Ecosystem Service.

- Make a distinction between functions, services and benefits (although there is no consensus on a classification).
- Comprise contrasting scenarios, establishing benefits and costs as a function of changes between alternative options.

Essentially, this approach involves extracting a defined element of an ecosystem, comprising one or more Ecosystem Services and modelling their behaviour in a number of different scenarios.

It is important to note that this assessment methodology does not involve itself with the ecosystem mechanisms or interactions which result in the services being considered.

This form of cost-benefit analysis leads to a ranking of alternative scenarios, in relation to their effects on the services provided. It does not seek to provide either an absolute valuation of the services being considered, or relate the area under assessment with any other.

This form of option ranking, in a qualitative manner, is a familiar approach in the field of Environmental Impact Assessment. Therefore, it would seamlessly fit in with current practice in terms of defining the option with the least effect on Ecosystem Services, provided sufficient information can be made available for the ranking to be objectively robust.

The second chapter 'Biodiversity, ecosystems and Ecosystem Services' includes something of a contradiction. Initially it asserts that all ecosystems are shaped by people and that people and ecosystems are interdependent social-ecological systems. This implies that people and ecosystems are separate. However, it then goes on to include people within the living component of the environment, which interacts with the non-living components to form an ecosystem.

Rather than being a holistic approach to the understanding of Ecosystem Services, it places human activity in a grey area which offers significant potential for confusion and misunderstanding when making assertions about ecosystems and the services they provide. Under this approach, the human involvement in the generation of Ecosystem Services through affecting ecological processes can be flexibly either included or excluded from consideration without any clear definition.

This chapter cites three main ways in which biological diversity relates to the operations of ecosystems and Ecosystem Services. These theoretical mechanisms, which are formulated as a manifesto for the essential role of biological diversity in the operation of Ecosystem Services, unfortunately have little support and are not logically integrated.

Firstly, it states that an increase in diversity often leads to an increase in productivity due to complementary traits amongst species for resource use, and productivity itself underpins many Ecosystem Services. This assertion is flawed in several areas, firstly an ecosystem is a self-regulating functional unit, and productivity (i.e. excess materials which can then be exploited) are not produced by any such system, all internal productivity is re-used internally by the ecosystem, when it is considered as a 'black box' functional unit. It is wrong to assume that increasing the diversity of any given unit will increase productivity, this implication cannot be supported, as each species makes use of specific resources for its own survival, there is no such internal co-ordination to allow a concept such as 'complimentary traits' to be assigned in any mechanistic manner. The phrasing of the assertion, implies that 'increase in diversity' is something that man has the ability to see as a desirable trait, and should in some way undertake interventions to achieve. This concept, of course is also contradicted by the historical interventions of man where individual species have been extracted from their natural ecosystems and exploited in isolation for their beneficial traits through cultivation.

Biological diversity is a feature which can, under present knowledge, only be considered in terms of energetic interactions between species. More complex systems have more complex interactions occurring within the 'black box' of the ecosystem under consideration leading to a system which maximises resource use. Little more can be asserted on this matter, and certainly nothing regarding productivity (output).

Secondly, this chapter asserts that increased diversity leads to an increase in response diversity (range of traits related to how species within the same functional group respond to environmental drivers) resulting in less variability in functioning over time as environment changes. Although this has a theoretical rationale [7], real-world evidence is not overwhelmingly conclusive that the causal link between biodiversity and stability is as clear cut as this chapter suggests [8].

The third assertion identifies certain species as being disproportionately important in the maintenance of ecosystems in terms of their provision of particular Ecosystem Services (i.e. species important to mankind's continued access to any given Ecosystem Service). Apart from contradicting the first assertion regarding biological diversity, this statement fails to acknowledge that we have such a limited understanding of the species present that it is not possible to identify any such 'keystone species' with any degree of confidence. As previously discussed, many of the large, charismatic vertebrates which are generally the focus of modern conservation projects are unlikely to be those aspects of biological diversity which should be concentrated upon, even though they receive the vast majority of funding for such works.

The central tenant of this chapter is that ecosystems should be managed, based upon existing understanding. The urgent need to develop more knowledge and understanding of ecosystems and their individual elements is noted, but not addressed.

The third chapter of this document 'Measuring biophysical quantities and the use of indicators' acknowledges a lack of knowledge in respect of the selection of indicators for biodiversity, and recognises that it has hampered the ability to undertake economic valuations and assessments.

One of the key shortfalls recognised is the inability of the existing indicators of biodiversity to be translated into human benefits via Ecosystem Services. However, it does not highlight the inadequacy of current indicators because they are incomplete, but because they are measuring the wrong thing in respect of Ecosystem Services.

This chapter adopts the stance that it is necessary to develop a set of indicators which should be monitored and modelled in order to provide metrics which can be convertible into economic terms. This approach is one which is accepted in mainstream economics as the way in which economic systems are understood and managed. One big problem with such an approach, however, is that by abstracting the nature of biodiversity and ecosystems through the use of indicators it presents a false sense of certainty. It must be borne in mind that even the best indicators can currently only be 10% accurate, as 90% of biological diversity (in terms of species) is literally unknown. The relevance of indicators to this 90% can only be purely speculative.

A further issue with the use of such indicators is that all the evidence from mainstream economics shows that once an indicator is formulated, all efforts are applied to its maintenance and management. Such indicators become a surrogate for reality, despite their inadequate ability to tell the whole story. The use of GDP is such an example, where individual well-being is a secondary consideration to the health of the indicator.

The fourth chapter 'Socio-cultural context of ecosystem and biodiversity valuation' frames the exercise of economic valuation in terms of cultural belief systems and as a social feedback mechanism, to allow relationships between society and the natural environment to be investigated.

Valuation, in this context, is seen to be a relative exercise. This brings to the fore the concept that some features of the natural environment are of greater value to some stakeholders than others. This, of course raises the question of issues of wealth and control of land and resources. Although some communities may place a great deal of value on a certain element, they may be 'outbid' by commercial interests with greater wealth negotiating directly with those who claim ownership of the resources under consideration.

In this chapter intrinsic values are defined as 'culturally embedded moral truths'. It asserts that such intrinsic values need to be taken into account by including the voices of institutionalised cultural representatives in the valuation process. By adopting such a procedure, however, there is the very real possibility that cultural change and adaptation will be stunted by the economic pressures of maintaining a particular cultural standpoint. Such effects result in the establishment of artificially maintained cultural/environmental 'theme parks', where the economic pressure to maintain a cultural purity outweighs motivations for individual innovation, growth and development.

Chapter five concerns itself with the economics of valuing Ecosystem Services and biodiversity. It commences by defining economics as: 'the study of how to allocate limited resources'. This is a definition which appears to be derived from Robbins [9] who states: 'Economics is the science which studies human behaviour as a relationship between ends and scarce means which have alternative uses'. The definition used in this chapter tends to skew the intention of the definition of economics towards state control and collectivised allocation of limited resources, rather than the non-interventionist, free-market approach of the original definition.

This chapter identifies a variety of the difficulties associated with economic valuation and uncertainties in the application of economic valuations to ecosystems and Ecosystem Services. It, however, fails to emphasise the lack of knowledge and understanding surrounding biodiversity itself.

Despite the difficulties and uncertainties surrounding the process, this chapter proceeds to present an outline approach to valuation with the intention of demonstrating that 'biodiversity and Ecosystem Services are scarce and that their depreciation or degradation has associated costs to society'.

There is an assumption in this chapter that human societies are not currently existing in a stable state alongside natural systems. The presupposition is that there is an ongoing and unrelenting degradation of natural systems and that it is the responsibility of policy makers and stakeholders to redress this situation through the recognition of the value of Ecosystem Services and adopting the responsibility to maintain them, at a financial cost, via the levying of taxation and other fiscal means.

The chapter fails to recognise that it is the prevailing mainstream economic theory which requires year on year growth in order to service national, corporate and private debt, together with the instigation of public works during times of economic downturns which are the main drivers of environmental degradation, rather than the day to day activities of ordinary citizens undertaking routine economic activities.

The proposed approach to valuation fails to draw Ecosystem Services into the wider economic arena. Rather, its remit is to present a stand-alone mechanism to allow for the management and conservation of natural systems to be funded.

It approaches valuation by adopting the Total Economic Value (TEV) framework. This is a form of cost-benefit analysis in which ecosystems are considered as producers of output values, which can be enumerated in terms of Ecosystem Services, together with additional value (option value) of services which are not currently being made use of, but may be required at some point in the future.

The report acknowledges that this approach is an uncertain one, due to insufficient knowledge of ecosystem dynamics, human preferences and technical issues associated with the valuation process. It also recognises the fact that the approach would only operate on a case by case basis, and would be unlikely to offer universally applicable results.

Chapter six, of the report is entitled 'Discounting, ethics, and options for maintaining biodiversity and ecosystem integrity'. It addresses the issue of how the use of biological resources now has the potential to affect the availability of such resources for future generations and how this can be managed in a controlled economy.

The use of a capital investment approach to the issue is dismissed as inappropriate in this case. As previously mentioned, the fiat monetary and banking system can cause distortions through the creation of debt at will. In this case the choices of whether to retain natural resources or use them and then 'bank' the subsequent income as a monetary gain is skewed almost entirely by prevailing financial interest rates. Due to the irreversible nature of environmental damage, the lack of knowledge and understanding that we have of the systems involved, and the long time frames involved, it is often considered that the mainstream economic model does not adequately deal with environmental issues [10].

Despite the inability of current economic theory to deliver valuations, the chapter states that aggressive conservation and ecosystem restoration policies will be required to sustain human welfare with regard to the support received from natural resources.

The process used to assess the relative current and future value of natural resources is through the calculation of a discount rate. In a business context, the discount rate is used to determine how much future cash flows of a business would be worth as a lump sum total today. In environmental terms the discount rate would be used to determine future Ecosystem Services and their value, presented as a cash value at the present time. In part, this approach is an attempt to address the previously identified issue relating to the inflationary requirements of fiat currency in mainstream economics, which means that a quantity of currency now will have less value in the future.

However, the calculation of a discount rate, although using mathematical formulae, cannot bring quantitative accuracy to the calculation due to the number of assumptions required. Therefore, qualitative issues such as ethics, best guesses about the well-being of those in future, and preserving life opportunities must be included in what eventually becomes an intangible figure. This, however does not deter the authors from presenting the varied uses and applications of discount rates in the formulation of policy decisions and how they may affect Ecosystem Services into the future.

For example, the chapter undertakes calculations which determine that a 5% discount rate implies that biodiversity loss 50 years from now will be valued at only 1/7 of the same amount of biodiversity loss today. This should be read in the context that our knowledge currently only extends to around 10% of biodiversity, and that biodiversity itself is not an Ecosystem Service.

Although the TEEB approach to valuing Ecosystem Services presents a wealth of information regarding the intricacies of undertaking the exercise of valuation, it loses focus in several key areas. Firstly, it takes such a broad and flexible definition of the terms ecosystems and biodiversity, that almost everything may be included within the analysis. By taking such an approach, the only relevant actionable examples presented cannot reveal any general principles and can only be used as individual isolated case studies, which fail to cast any insight onto more general mechanisms, or indeed other individual cases. Secondly, it is entirely embedded within the current mainstream economic concept of a managed economy, and makes a pre-supposition that the management of the natural environment is always an essential part of the delivery of the benefits of Ecosystem Services.

This aspect of mainstream economics is retained as an article of dogma (interventionist actions to achieve a theoretically optimal outcome), whereas the report is happy to admit that mainstream economics is inadequate for valuation purposes. The problem with such a situation is that it fragments the approach to the natural environment by only refuting aspects of mainstream economics which clearly fail to operate, whilst doggedly maintaining other ideological aspects without noting that they are all part of an overarching economic structure. This leads to the theoretical elaboration of certain elements, such as valuation in order to allow overall policy to continue to be driven by mainstream economic ideas. This situation is reminiscent of the Ptolemaic elaboration of the geocentric solar system, adding complexity to maintain dogma.

At this stage, it is worthwhile stepping back from the TEEB approach to Ecosystem Services, and look at ecosystems from first principles in order to try and develop alternative perspectives in relation to the interrelationships between mankind and the environment.

Ecosystems as a Consumer

The concept of Ecosystem Services emphasises the idea of the environment as a producer of goods and services. However, if we are to take this view, then it is also necessary to investigate the flip side of the coin. That is by viewing ecosystems as a consumer of human resources when it comes to the management, conservation and maintenance of habitats, ecosystems and protected species. Although this input of human resources can be framed as an investment in Ecosystem Services, few causal links have been made which categorically prove that these activities are directed efficiently, or even appropriately, for the described purpose. Once again, the main issue in this regard is our lack of a complete picture of how such systems operate in the absence of human intervention. Due to this, such activities, although instigated with the best of intentions, have a chance of causing more harm than good, either directly through ignorance of the full implications of the actions taken, or indirectly through a misallocation of resources and energy.

This issue is of particular importance because the use of Ecosystem Services as a vehicle for policy decisions at governmental levels can induce spending commitments of public funds and incur additional costs on businesses without necessarily providing a desirable outcome.

For example, a recent newspaper report [11] describes the translocation of a colony of 40 Great Crested Newts at a cost of £40,000 prior to a development on a site which is already industrialised. This is a good example of the scale of resource consumption which can surround the conservation of legally protected species.

Great Crested Newts (*Triturus cristatus*) have the lowest level of threat (Least Concern) under international IUCN guidelines [12], but are afforded the highest level of protection within the European Union. In the UK however, this is a widespread species (Fig. 6) and is found in approximately 75,000 localities.

Whether the expense associated with Great Crested Newts in the UK is considered to be critical conservation or a misallocation of funds is a matter for discussion. However, it is an undeniable fact that this species is a net consumer of resources, as there are no identified Ecosystem Services associated with this individual species.

Viewing ecosystems as consumers is a necessary, although sometimes uncomfortable, counterpoint to the view that Ecosystem Services are provided without human input. The extent to which our spending on the environment needs to be included in any cost-benefit analysis of Ecosystem Services is something which has not been directly addressed. It raises important questions regarding funding for conservation of species which do not provide any tangible service, and represent only a single species focus, which does not necessarily benefit underlying biodiversity in a demonstrable manner.

Projects which fail to provide the desired outcomes (assuming such outcomes are correctly framed) is analogous to the Austrian view of malinvestment which takes place during times of rapid economic growth. During phases of heightened environmental concern, it is relatively easy for funding to be applied to projects as a kneejerk reaction to the latest trends and news headlines without undertaking due diligence regarding the outcomes of such work. Similarly, legislative restrictions can be applied according to the strength of lobbying and popular opinion, irrespective of outcomes.

If we are truly experiencing such a boom in funding, the necessary question which follows is what happens when the output of poorly planned and framed projects and legislation fail to meet their objectives? This would be the ecological equivalent of a recession. In mainstream economics, such an eventuality would be addressed by increasing public funding and environmental protection measures, the Austrian school of economics would advocate allowing projects which have failed to meet their objectives to cease, and the re-direction of funding into more

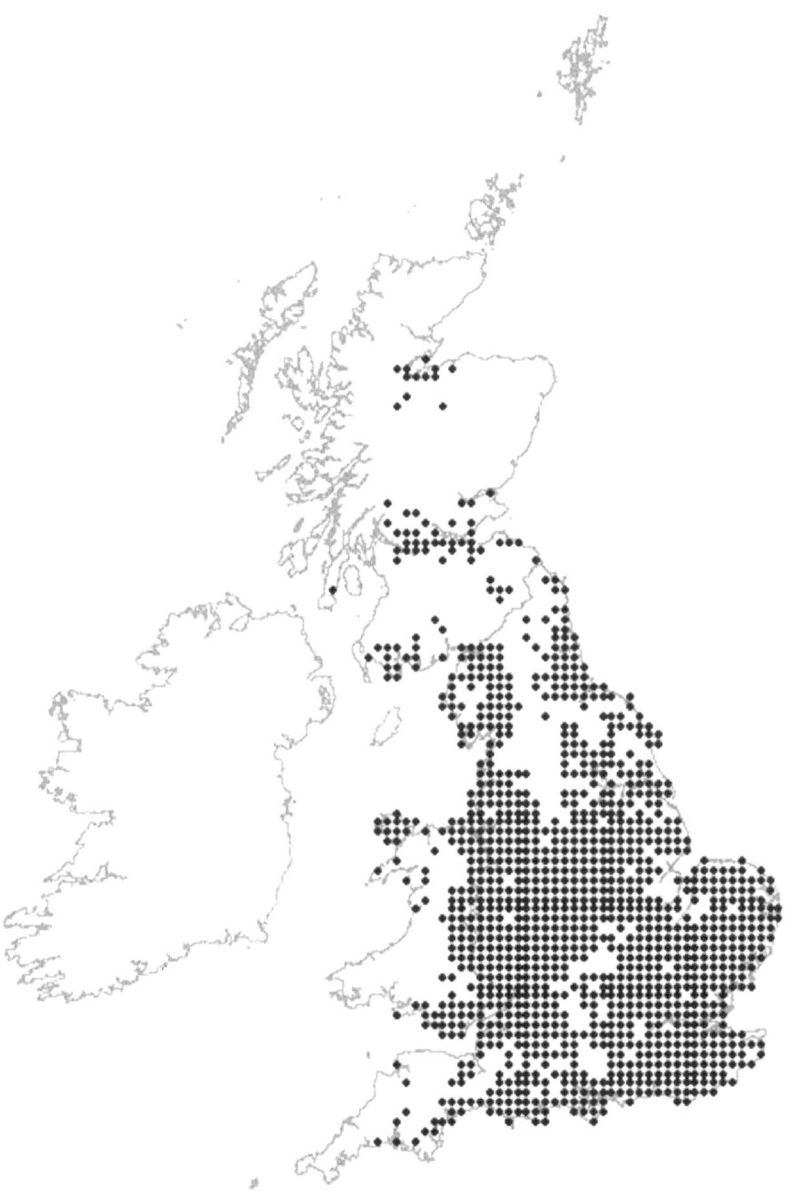

Fig. 6   UK distribution of Great Crested Newt (*Source* JNCC)

(ecologically) productive enterprises. Of course, our ability to determine the success or failure of an environmental project, or the success of a certain series of legislative protection measures is so poorly formed that it is rarely stated that a particular effort is simply not working, or if working, its benefits are minimal outside of a circular self-fulfilling set of aims.

It is a very difficult proposition, both practically and politically, to 'unprotect' a species which has received a great deal of funding and then reallocate funds to a different project. Rather, the mainstream approach is to keep pumping money into such projects despite minimal ecological returns on the investment.

Despite the different viewpoints regarding ecosystems and its underlying biodiversity, the lack of an understanding of the workings of the systems involved represents a major obstacle in identifying how best funds and investment can be allocated to maximise ecological returns as opposed to being seen to be 'green' in a rather superficial, demagogic sense.

There doesn't appear to be a means (in Western thought at least) of achieving a truly laissez-faire approach to the natural environment, where the natural productivity of the land (in respect of goods and services) is actually repaid through the non-intervention of mankind. It seems the desire to control, manage and actively conserve is an almost overwhelming impulse. The recent movement which seeks to 'rewild' natural areas is essentially non-interventionist in nature. However, as this movement gathers greater support, there is likely to be an upsurge in such areas becoming targeted as 'ecotourist' honeypots. This will, counterproductively, require inputs to direct and control human traffic, limit the excesses of human behaviour in the natural environment and provide a range of additional facilities. High profile non-intervention will ultimately, therefore, fall victim to its own success and require interventions of its own.

It appears that the role of the environment as a consumer of human resources is here to stay. Not that this is necessarily negative, provided that our activities are well focussed, with predictable outcomes and represent a positive net gain for both the producer of the goods and services (the environment) and the investor (mankind). Such equations, resulting in win-win situations are commonplace in economics, and are considered essential in a free market, so there is no reason why such transactions may not occur between mankind and our environment.

Ecosystems as a Commodity

At a basic level we can consider the natural (living) environment as a provider of ecosystem goods AND services. It is important at this point to define more clearly that we are speaking of elements of ecosystems which have a distinct and definite biotic component in their provision. Such provision should be direct and measurable. In this way issues such as weather, tides, climate, sea levels, global temperature, topography and mineral deposits are removed from consideration (these really should be considered as 'geophysical services', a subject which is equally worthy of study, but outside of the scope of this work). This allows considerably more progress to be made in our understanding, as the focus of consideration can be more finely tuned. As the archery saying goes: 'if you aim small, you miss small'.

If we are to consider aspects of the natural world as commodities then we need to provide an accurate definition of the term. In economics, a commodity is a good or service which is traded as an undifferentiated item. In other words, there can be no preference in terms of type within the commodity as a whole. Therefore, cars cannot be considered a commodity, each model is different in terms of brand, performance, price and desirability. Alternatively, a refined metal such as copper is treated as a commodity in that its origin, the miner and refiner involved are unimportant when considering the price to be paid.

As in all things, though, there is a spectrum within what may or may not be considered a commodity. For example tea is a commodity, but there is a variation in the types of tea available and some are much more expensive than others. In similar vein organic produce typically has a premium on price due to a combination of potential for lower yields and a fashionable desire for such products, which gives organic foods a greater demand in the marketplace.

When it comes to the goods and services provided by the natural environment one must look at the definition of commodity, to see if they fit and to what extent. The point of assigning the status of a commodity to the goods and services derived from natural world is that qualifying elements would then become considered as a universally tradeable item with a defined economic value. It also allows one to identify the areas where such goods and services enter the chain which links raw commodities with finished consumer goods. This analytical approach would clearly highlight to all concerned the value and economic importance of the goods and services concerned.

The clearest case for considering Ecosystem Services as commodities comes with the provisioning services. These are most definitely things which tangibly enter the marketplace (food sources, ingredients and raw materials). The commoditization of such Ecosystem Services is, however, more likely to result in their domestication and transference into the agricultural sector than maintaining them in a natural state. This is particularly true if demand for such products is high.

Other forms of Ecosystem Services tend to have a localised operating area and are essentially immobile, such as flood relief in a particular catchment or prevention of soil erosion in an area of agricultural activity. Although these services cannot be considered as commodities in the sense of the definition above, as they are not tradeable as such in their own right, they do have an economic impact beyond their direct effect in that they reduce risk. Such natural risk reduction features do actually have a form of commodity value in terms of insurance. If one considers insurance to be a commodity, in that it represents an undifferentiated and tradeable service (despite protestations from individual risk assessment and insurance businesses, who wish to remain as differentiated as possible), the concept of regulating services being a part of this commodity trade is not an unrealistic viewpoint.

Other Ecosystem Services, which are associated with cultural aspects are clearly not capable of consideration as a commodity. It is not possible to trade ones moral and ethical values, religious or spiritual beliefs, either as an individual or as a society.

Some schools of thought assign human labour as the primary economic commodity [13], which, when applied to material substances and objects, produces the goods and services that are traded in order to meet mankind's needs and desires. Using this interpretation of commodities, copper is valued due to the human labour expended in prospecting mining and refining (or recycling) it. This approach would exclude all natural systems which passively provide goods and services from consideration as commodities, although provisioning services could still be framed as commodities, as human labour is involved in their extraction, transportation etc.

So, from an economic viewpoint one can look at Ecosystem Services as falling into three categories: having value as a commodity, having value as a risk-reducing asset (insurance commodity) or having value as an ethical, spiritual or religious asset. Therefore, it is possible to split out some aspects of Ecosystem Services and allow value to be determined by trading on the open market. There is no need for exceptionalism in this respect as they are as subject to supply and demand economics as much as any other element of the economy.

It is worth noting, though, that the theoretical framework upon which mainstream economics is based is not necessarily the most benevolent approach to such goods and services, which would be subject to speculative trading, use as leverage for debt and being packaged into exchange-traded funds on the derivatives market. There is, therefore a clear danger not only of anonymising Ecosystem Services by their commoditisation but their inadvertent destruction during times of market fluctuations. So, mainstream markets do not provide a safe place for Ecosystem Services, but there is no objective justification for their exclusion from the market. This presents an apparent impasse.

Biodiversity is universally accepted as the foundation for ecosystems and their services. It is excluded from any attempt at valuation, primarily because our demonstrable lack of knowledge regarding species at a global level precludes any attempt at quantitative analysis.

Notwithstanding that issue, it is still possible to take the broad concept of biodiversity to asses its potential fit as a commodity. In the way that biodiversity is characterised in literature relating to Ecosystem Services, it does fall into a broad category which is undifferentiated. It is not possible to qualitatively distinguish biodiversity loss in the arctic from biodiversity loss in the tropics. Furthermore, biodiversity is described as the foundation (or 'raw material') from which Ecosystem Services are derived. If we consider all such Ecosystem Services as derivatives of biodiversity, then, in aggregate, it represents the underlying biotic commodity of all ecosystems and their services (including those which are not susceptible to economic valuation).

In functional terms, we have shown that biodiversity can be considered compliant with the definition of a commodity. It is at this point however, that we reach a major snag. Our knowledge of this apparent commodity is so limited at the present time that it would be impossible to make it tradeable in an open market as no one would be fully cognisant of what was being bought and sold. This places biodiversity in the unique position of being an untradeable commodity. Figure 7 shows a potential reclassification of Ecosystem Services, based upon the concept of commodity value.

Even though we cannot currently make use of biodiversity as a means of valuing the natural environment via the means of discovering its tradeable value, the ability to classify biodiversity as a commodity opens up some interesting prospects with regard to the valuation of the environment. These will be explored further in chapter "A New Model".

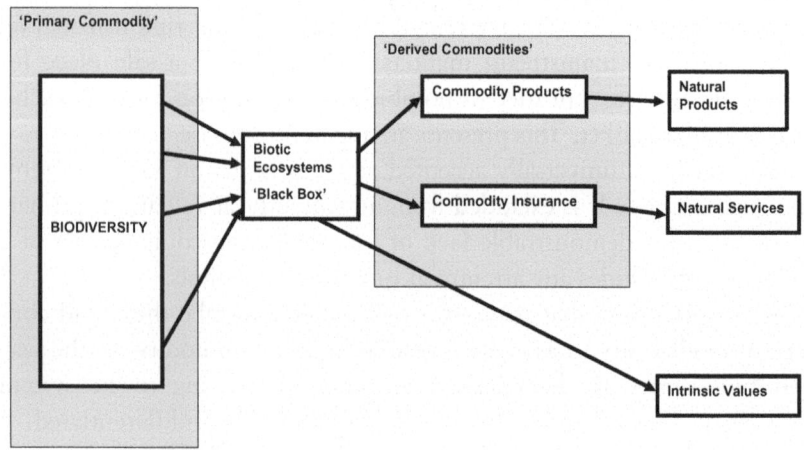

Fig. 7 Commodity based model of Ecosystem Services

Ecosystems as Knowledge

The theme underlying much of the discussion so far has been the lack of knowledge that humanity has with respect to natural systems. Despite this undeniable fact, we continue to behave as though we have some insight into the core concepts of 'biodiversity' and 'ecosystems' which allows us to effectively manage, restore and otherwise intervene in these systems. What is missing is any attempt at achieving greater insight into the results of our actions. As long as the intentions are good then no blame is placed when outcomes are less than desirable.

Imagine a situation where the world operates upon the knowledge held within an encyclopaedia. In this world there are experts who claim 'encyclopaedic knowledge' and are held in high regard, policies and legislation are framed on the basis of the knowledge which they hold. However, consider the disappointment which one would feel when one discovers that this encyclopaedic knowledge is in terms of the number of volumes in the encyclopaedia, the size, weight and form of each book, the type of writing on the spine and the materials from which the encyclopaedia is made. There is no one who has actually opened the book and read what is inside. Such efforts are dismissed as too time-consuming, too costly etc. The principle which holds sway in the corridors of power is: 'as long as we know the encyclopaedias are there and the superficial form they take then we have sufficient information to act in an informed manner'.

Although an extreme analogy, this represents something of our current state of knowledge about biodiversity.

> It is difficult and often even impossible to characterise the functioning of a complex system, such as an eco-agrosystem, by means of direct measurements. The size of the system, the complexity of the interactions involved, or the difficulty and cost of the measurements needed are often crippling. [14]

It is unusual, to say the least, for mankind to adopt such a defeatist attitude when faced with the boundaries of knowledge. Take for example the study of astronomy, where resources and effort is directed towards enumerating and identifying the stars and planets of other solar systems, and even other galaxies.

> The cosmos is all that is or ever was or ever will be. Our feeblest contemplations of the Cosmos stir us—there is a tingling in the spine, a catch in the voice, a faint sensation, as if a distant memory, or falling from a height. We know we are approaching the greatest of mysteries. Carl Sagan

> To confine our attention to terrestrial matters would be to limit the human spirit. Stephen Hawking.

This contrast in thirst for understanding is something which is worthy of consideration. What is it about the environment which is such an overwhelming deterrent to our inherent curiosity?

The stars are remote, we cannot affect them and they cannot affect us in a meaningful manner (assuming the probability of alien invasion is negligible). Due to this 'effects barrier' it is possible to wax lyrical about space with the full confidence that there can be no disagreement and no hypocritical disjunct between words and actions.

The situation which emerges between man and the environment is very different, we can and do affect the natural world. There is a risk, which is magnified at the level of national economic management, that full knowledge of our actions would result in an effective brake on what is termed 'economic growth'. Anything which could affect this mainstream concept of growth is seen as anathema by mainstream economists and politicians. Therefore, there is a top down desire to 'let sleeping dogs lie', when it comes to revealing the true nature of biodiversity and ecosystems. This is easily achieved under the current system, funding for basic taxonomic and ecological studies is limited at the top, therefore choking off activity.

Ignorance is bliss, or is it?

There is an alternative view of knowledge. It could be considered to provide a source of value. This is recognised in the pharmaceutical arena, where natural systems are effectively searched for potentially useful chemical compounds of biological origin. Of course, this work is not conducted in the public realm for reasons of intellectual property. Such work offers the potential for pharmaceutical companies to effectively take ownership of valuable biodiversity resources in order to provide them with the recompense required for prospecting, isolating and testing potential new drugs.

Ecosystems and their component species and habitats represent a massive potential source of knowledge and understanding. However, mainstream economics and the politically driven financial system does not recognise that the investigation of these resources has the potential to be a legitimate economic activity. Furthermore, there is the distinct possibility that its neglect is willful in nature, in order to allow guilt-free economic growth to prevail.

We have already determined that biodiversity has the characteristics which would qualify it as a commodity. Following that reasoning, the concept of biodiversity as knowledge represents the prospecting, extracting and refining the commodity to allow the corpus of biodiversity knowledge to be brought into a marketable state (alternatively considered as a state which permits value to be recognised in the economic arena).

References

1. Troudet, J., Grandcolas, P., Blin, A., Vignes-Lebbe, R., & Legendre, F. (2017). Taxonomic Bias in Biodiversity Data and Societal Preferences. *Nature: Scientific Reports, 7*(1), 9132.
2. Spash, C. L. (2015). Bulldozing Biodiversity: The Economics of Offsets and Trading in Nature. *Biological Conservation, 192,* 541–551.
3. Batavia, C., & Nelson, M. P. (2017). For Goodness Sake! What Is Intrinsic Value and Why Should We Care? *Biological Conservation, 209,* 366–376.
4. Piccolo, J. J. (2017). Intrinsic Values in Nature: Objective Good or Simply Half of an Unhelpful Dichotomy? *Journal for Nature Conservation, 37,* 8–11.
5. Mises, L. von. (1949). *Human Action.* New Haven: Yale University Press.
6. Kumar, P. (Ed.). (2010). *The Economics of Ecosystems and Biodiversity Ecological and Economic Foundations.* London: Earthscan.
7. Doak, D. F., Bigger, D., Harding, M. K., Marvier, M. A., O'Malley, R. E., & Thomson, D. (1998). The Statistical Inevitability of Stability-Diversity

Relationships in Community Ecology. *The American Naturalist, 151*(3), 264–276.

8. Worm, B., & Duffy, J. E. (2003). Biodiversity, Productivity and Stability in Real Food Webs. *Trends in Ecology & Evolution, 18*(12), 628–632.

9. Robbins, L. (1932). *An Essay on the Nature and Significance of Economic Science.* London: Macmillan.

10. Dasgupta, P. (2008). Nature in Economics. *Environmental & Resource Economics, 39,* 1–7.

11. Pittam, D. (2018). Great Crested Newts at Site of Multi-Million Pound Building Project Will Cost Up to £40,000 to Move. *Nottingham Post.*

12. iucnredlist.org.

13. Marx, K. (1867). *Capital, Volume I.* New York: International Publishers, 1967.

14. Girardin, P., Bockstaller, C., & Van der Werf, H. (1999). Indicators: Tools to Evaluate the Environmental Impacts of Farming Systems. *Journal of Sustainable Agriculture, 13*(4), 5–21.

A New Model

The previous chapters of this book have demonstrated that the current concept of Ecosystem Services, when used as a means of applying value to the environmental benefits mankind receives from natural or 'unmanaged' areas, can only truly be applied under conditions where there is some form of fixed basis for the valuations carried out. Without such a basis, any application of a monetary scale is only established in a relative way. This is due to the fiat nature of all currencies, which means that geopolitical issues can have effects on an economy that could render an entire nations' environmental wealth (as expressed through Ecosystem Services and valued in fiat currencies) valueless.

The current mainstream economic system is based upon the principles of promoting consumption (demand) to drive economic growth. This approach leads to a competition for growth amongst different nations, generally quantified in terms of increase in GDP. This principle leads inevitably to the idea that an economy can grow more rapidly than its rivals when debt is introduced into the system, by incurring debt it is believed that it is possible to stimulate more growth within the economy. The obvious problem with this approach is that debt must at some point be repaid, together with ongoing interest payments for

© The Author(s) 2019
S. Muddiman, *Ecosystem Services*,
Palgrave Studies in Natural Resource Management,
https://doi.org/10.1007/978-3-030-13819-6_5

the loan. This awkward situation is dealt with via mechanisms which maintain a level of inflation in an economy. In this way the debt becomes managed through the effective devaluation of the currency on a year by year basis. Of course, the side effect of this is that the value of a currency in terms of its purchasing power will decline over time. It is clear that this model does not provide a stable or fixed value for any commodity or resource over time, including valuations applied to Ecosystem Services.

In order for any such valuation to keep up with the credit and debt driven need to grow a business or a national economy, it is clearly necessary for such valuations to be incorporated into some form of marketplace, so that they are able to alter and appreciate in value, at least in line with the prevailing rate of inflation at any given moment in time. However, the present framework of 'markets' and financial instruments would actually present an additional handicap to the valuation of Ecosystem Services. Valuations would immediately be distorted through not only speculative investments and the effects of non-related market forces, but also through the inevitable market for Exchange Traded Funds (a marketable security that tracks a stock index, a commodity, bonds, or a basket of assets) and other derivative products [1], where investors would seek either exposure to the market without direct involvement, or would be betting on changes in value (either higher or lower). Such mechanisms would lead to a situation where the larger financial commitment in derivatives would effectively control the value of Ecosystem Services themselves (the tail would be wagging the dog).

Additionally, by throwing Ecosystem Services into mainstream finance, the environment inevitably becomes even more remote from its most closely associated communities and increasingly the domain of governmental and financial sector interests. There is a grave danger in these circumstances that the stakeholders on the ground, who have direct dependence upon Ecosystem Services, will have an even more diminished say in the persistence of their lifestyle.

For example, as Ecosystem Services become financialized, there are possibilities that at some point in the future a calculation relating to the cost of an engineered flood relief scheme could be seen as outweighing an existing, natural system. This is particularly likely when calculations adopt a mainstream economic view of government intervention. This approach would be to enhance employment during times of an economic downturn. Under these conditions a temporary and speculative economic stimulus (driven by Government borrowing) could easily lead to the permanent loss of an environmental system which offers far more in terms of biodiversity than its Ecosystem function, this biodiversity knowledge would be permanently lost upon its destruction.

The above example leads us to another issue relating to the inevitable financialization of Ecosystem Services. Under the situation of an area having no classifiable or marketable Ecosystem Service value, its value in terms of such services is essentially considered to be zero. However, this focus neglects to consider the unrecognised knowledge of its unrecorded biodiversity. Such areas could be destroyed and developed without any need to make a cost-benefit assessment in economic terms.

By concentrating on the value of Ecosystem Services rather than its foundational biodiversity there are real issues in terms of overlooking the potential of elements we currently have no knowledge of. As unrecognised biodiversity cannot be valued, and no effort is made to fully extract the biodiversity knowledge of an area, there can be no conclusions drawn regarding what is being lost. Therefore, there is a strong element of 'ignorance is bliss' in the adoption of Ecosystem Service valuation. The idea of working with a full data set is dismissed as being unrealistic, impossible or even more tellingly, too expensive.

That having been stated, the development of the concept of Ecosystem Services is undoubtedly a valuable tool to allow environmental issues to be incorporated into the sphere of human activity. However, it should be looked at critically when considering how modern financing of potentially environmental damaging growth and development takes place.

An Alternative Framework

The current application of Ecosystem Services tends to put the cart before the horse. It assumes that we have sufficient information to be confident that the observed services which are received from ecosystems have been sufficiently characterised and that it is the services which are the key factor to be considered when valuing ecosystems for making informed policy decisions. In addition, there is a general view that Ecosystem Services require a constant financial input to manage/control them. The alternative view is that non-intervention approaches to the natural environment are the optimal means of ensuring their survival. The best approach for mankind to have to the environment is that of an observer and recorder of diversity.

Interventions should be considered only as a last resort, rather than the first port of call. This resonates with the Austrian view of the economy, where fiscal and monetary intervention is perceived as doing more harm than good.

The framework proposed below involves the application of four principles which place a different perspective on the nature of Ecosystem Services which:

- Places limits on credit for growth affecting the natural environment.
- Places the community and individuals who derive the most benefit from Ecosystem Services in control of their future.
- Recognises the true nature of sound money.
- Applies the concept of valuing biodiversity knowledge.

Credit

Much of the activity humanity undertakes is currently not financed by the result of productive activity via savings and profits, the majority is based upon incurring debt and obtaining credit, with the assumption that future commercial profits or national economic growth will provide the necessary revenue for repayment (together with any interest accrued). Due to this, the application of a system which superimposes the reality of useful natural services (both recognised and as yet undetermined) onto a system where value is very much an elastic phenomenon should be treated with a great deal of caution.

The principles of the Austrian School of economics provides a theoretical framework by which the valuation of Ecosystem Services can be carried out in a way which permits a more realistic result in terms of providing a solid foundation for economic cost benefit analysis to be undertaken. Although it should be noted that such analytical and mathematical approaches are often considered anathema to Austrian School economists, as they assert that human nature and human action are not suitable subjects for mathematical quantification [2].

One of the features of Austrian Economics is the imposition of self-imposed limits on the extent by which credit can be used for developments, and allows for valuations to be far more stable over time, as they are established in accordance with a fixed (non-fiat) money supply, rather than allowing value to be 'inflated away' through monetary expansion.

Changing the entire economic system in order to accommodate the use of a stable valuation for Ecosystem Services is clearly not a viable option, but it is possible to develop a system which accommodates both approaches under specific circumstances.

Such a 'two-speed' approach would give mainstream economics free reign to operate in areas where environmental damage was minimal. Such fields of economic activity would primarily involve sectors engaged in the 'regeneration, reuse and recycling' of land and materials. Credit and state financing of such activities would be maintained or enhanced in order that humanity could make the most efficient use possible of the resources which have already been extracted from the environment.

However, the credit and financing available for projects which affect any known, or as yet unrecognised, biodiversity elements which are outside of the managed environment should be strictly curtailed. The insistence that money obtained for such projects is already realised from other profitable activity will lead to a great deal more circumspection when embarking upon speculative projects which have a component of environmental damage. Those responsible for such projects would actually have some 'skin in the game'.

This approach would remove the ability to use financial leverage in order to stimulate growth at the expense of the natural environment. If it was necessary for proven profits to be used to fund future expansion into arenas affecting the natural world there would be a significant incentive for companies to make full use of existing resources. Furthermore, as profits are routinely returned to investors via share dividends, the onus would be upon corporate managers to convince these investors that the medium and long term profitability of undertaking activities which affect the environment is worth the short term sacrifice in dividend payments. This, in turn, would engage shareholders and induce a more and informed active role in the companies they are invested in.

Under the proposed framework, any project which has the potential to adversely affect biodiversity should be initially assumed to be not available for either credit or speculative investment. This will divorce the natural environment and its associated biodiversity and Ecosystem Services from the speculative need for national and corporate growth which is the hallmark of modern mainstream economics.

So, how should such projects be financed? In a world where debt-based development is not considered to be compatible with loss of Ecosystem Services, such projects would need to be funded from the revenues received through profit generating activities. This is one of the keystone ideas of sound economics, and it would require a significant change in both the speed and focus of economic growth.

The profits from existing activities would represent an appropriate measure of whether an actor (government or business) could afford to affect Ecosystem Services. Such profits would need to be derived from existing resources. This would moderate the mainstream economic view of public spending and leveraged private sector growth. Decisions would have to be made in a far longer timescale, and returns could not involve speculation on a resource or commodity being at its peak price to be rapidly exploited and then discarded. Developments involving high levels of risk would be far more cautiously approached and situations where a project is started, the environmental costs incurred and then abandoned would be far less likely to occur. Government spending would need to be based upon resources saved during the periods of strong growth, or focussed upon aspects such as recycling and reuse of materials. Regeneration and repair of infrastructure would become a more economically viable alternative to the continuous expansion into undeveloped areas.

This proposal represents the minimum application of sound financial principles. Although the money in this case remains 'fiat' and not based upon any underlying commodity, resource or real-world foundation. The insistence that financing is put in place in real terms to fund projects with potential environmentally detrimental effects offers a step towards.

Land Ownership and Property Rights

One key issue in relation to the development of Ecosystem Services as a subject of study which has not been fully resolved is the question of the ownership of the services under consideration. Although it is tempting to assign ownership to 'the planet earth' or other similarly loose concepts, the formulation of sound economics and policy decisions regarding Ecosystem Services requires the allocation of such services to a human stakeholder. There is frequently a mismatch between those who bear responsibility for the maintenance of an Ecosystem Service and those who benefit from it. This tends to require transfers of responsibility, funding and taxation in order to present a system which represents a sustainable, fair and equitable distribution of costs and benefits across all the stakeholders involved. Of course, this system inevitably involves legislative intervention and government/state involvement in all aspects of the process.

In addition, the question of security of ownership of an area of land which receives a certain Ecosystem Service is of critical importance. A lack of security in ownership can enable the entire concept of Ecosystem Services to be circumvented through mechanisms of compulsory relocation, compulsory purchase or other appropriation of land. It is important to recognise that the value of an Ecosystem Service is related only to its use. If a potential Ecosystem Service is recognised, but has no one who receives benefits from it, then this lack of recipient effectively negates the value of the service. Therefore, the removal of service users will take the value of the Ecosystem Service to zero, irrespective of the underlying biodiversity.

For example, if a settlement is protected from flooding by the presence of woodlands in the watershed, the woodlands have a clear value in this regard. However, if a proposed infrastructure development was proposed which according to mainstream economic theory would provide a boost to GDP and required compulsory resettlement, then the woodlands would no longer be considered of value in terms of flood protection, as they would not be providing this service to a settlement. This would then enhance their value for timber and commercial interests could exploit the now 'redundant' woodland without having to take account of the value of the Ecosystem Services they provided. This change in value bears no relation to the (unknown) biodiversity value of the woodland.

The issue of property rights is something which has been presented as an important aspect with regard to governance and policy issues [3]. The TEEB report considers there to be at least three arguments which support consideration of property rights and distributional impacts as an integral part of policy development:

- reasons of equity: fairness in addressing changes of rights between individuals, groups, communities and even generations is an important policy goal in most countries;
- taking distributional issues into account makes it much more feasible to achieve other goals when addressing biodiversity loss, particularly related to poverty alleviation and the Millennium Development Goals; and
- there are almost always winners and losers from policy change and in most cases, loser groups will oppose the policy measures. If distributional aspects are considered when designing policies, the chances of successful implementation can be improved.

The concept of property and ownership is essentially a human construct which is universally recognised within a legislative framework. However, the idea of 'ownership' includes a wide range of different systems.

Property can be owned by a community, allowing members to freely make use of the communal resource without requiring payment for qualifying users. This approach can lead to issues of the repair, maintenance and management of such property as such freely accessible rights do not incorporate obligations for either payment for upkeep or responsible use. In a small tight-knit society these problems can be overcome at an individual level, but as communal property reaches a larger scale, then opportunities to exploit the system become increasingly frequent. The axiom that everyone is responsible for a given task can rapidly degenerate into the idea that no-one is responsible.

One response to such issues is the establishment of a level of ownership by state or government. Under these circumstances, responsibilities are given to state officials to manage the area, whilst users of the land are granted rights and obligations with respect to the use of the land. Breach of obligations results in prosecution under the law. In order for this system to operate it is necessary for the state or government to fund the officials required, and furnish them with the resources necessary to execute their task. In addition, a legal framework is required to enforce obligations on the users of the property. Funding this system required a centrally raised revenue stream, which is typically some form of taxation. Such a system can lead to resentment, when those who pay the taxes do not consider themselves to be in receipt of proportional benefits. This approach also weakened by fluctuations in the economic cycle, which could result in the neglect of lands in public ownership during times of financial contraction, when funding is withdrawn and reallocated according to other priorities.

A third approach is the development of the concept of private property, which is backed up by legislative protection. Such a system where tools, equipment, land and homes are owned privately and maintained by the owners, who have sole rights to the benefits accrued from the use of the property is the most typical approach to ownership in the developed world. However, this model does present a number of difficulties in developing countries and when governmental or commercial interests clash with the ownership rights of individuals.

Where there is a lack of capital available to purchase the necessary materials to make one's property productive (such as machinery, seed, raw materials etc.) then interventions through either state subsidy or private sector funding can act to bridge that gap, although the debt repayments (plus interest) act to restrict the economic freedom of the borrowers, who become dependent upon the financing body for continued support until debts are paid off. This situation, which is well known for all holders of mortgages for property, are entrained into the prevailing economic system, and any attempt to deviate from it leads to confiscation (repossession) of property. One may be in possession of productive land, but that productivity must, at least in part, be converted into acceptable financial instruments for the lender.

In cases where there is seen to be an overwhelming benefit to a nation by undertaking infrastructure, resource extraction or other developments, most governments reserve the right to purchase or otherwise appropriate property for the purpose. As noted previously, this approach can inadvertently circumvent the value of Ecosystem Services, leading to opportunities for commercial exploitation.

One of the significant issues relating to ownership, property and Ecosystem Services is the recourse a property owner may have when a given service is affected by a third party remotely from the property. This has some parallels to the nuisance caused by noise, air pollution or other offsite effects.

The use of legal frameworks involving prosecution of transgressors are the means by which disputes resulting in an infringement of property rights can be dealt with. It is important to note that the law does not have the ability restore the status quo after an offence has been committed. It can, in the first instance act as a deterrent for certain acts, and in its second role it can afford blame and extract punishment. If for example, an area of land is damaged or destroyed by a third party, the law cannot restore the damage done, but can only monetise the damage and present an alternative restitution through a financial substitute (i.e. a fine) and provide some form of punitive action against the perpetrator (imprisonment, additional financial sanctions etc.).

Where land is unowned, there is no clear victim in respect of property, and therefore the concept of restitution is redundant, as there is no recipient of compensation. Under these circumstances the law becomes a preventative instrument only.

Until now, the main use of legal recourse has been in respect of the occurrence of nuisance to one's property by offsite influences, such as atmospheric pollution and noise. Although this could be used as a framework for dealing with effects on Ecosystem Services instigated by third parties offsite from the location of the service user, there remain a number of clear differences in the two situations. It will therefore be increasingly necessary to provide a legislative framework which accommodates issues of interference with Ecosystem Services.

One of the main differences is that the influence of atmospheric pollution and noise represent the result of a clearly measurable deterioration in the quality of an environmental baseline. These deteriorations can be measured quantitatively, and thresholds imposed by legislators. Under circumstances where these thresholds are breached it is possible to restore noise levels or air quality to a legally acceptable level. In the case of activities which affect Ecosystem Services, the options for remedial action are far less clear. The empirical measurement of an Ecosystem Service is much more difficult to accomplish than the straightforward chemical or biotic methods used when considering pollution levels. For example, answering a question such as 'to what extent has removal of tree coverage increased the likelihood of flood events' is unlikely to provide an answer which would permit a percentage threshold of tree cover to be established. These practical difficulties are compounded by the additional vagueness of our knowledge of biodiversity and its connection to many Ecosystem Services. For example, the biology and population dynamics of pollinators, their predators, parasites and full habitat requirements are so poorly known that any thresholds must be based upon gross simplifications of the actual situation, which may only be valid under certain, rarely experienced, circumstances.

This difficulty in establishing legally defensible thresholds for habitat loss and/or deterioration leads to an even more complex issue of how any breach of such thresholds can be remediated. Questions of timescale, monitoring and what actually constitutes a remediated situation (functionally speaking) are extremely difficult, if not impossible to establish in a manner which would allow proof of remediation to be demonstrated.

In addition to the problems presented by restoring Ecosystem Services, and measuring them, the other key contrast between Ecosystem Services and pollution issues is that although polluting events may represent a nuisance, their influence is not often primarily economic in nature. But the loss of Ecosystem Services, although they may be considered 'free', can have significant economic impacts upon the beneficiaries.

This presents a number of legal issues, which need to be resolved when considering Ecosystem Services. For example, if an owner of a property makes a decision to change his agricultural system which leads to a decrease in suitable habitat for pollinating insects, and this affects a neighbouring landowner who relies on insect pollinated crops, does the legal framework require the first landowner to cease from his plans, which may be considered to be an infringement of his right to farm his land as he sees fit is the second farmer provided with compensation from the first for loss of harvest? If so, then for how long? Or is the second farmer forced to change his agricultural practices also, to eliminate the need for the lost service, in which case does he receive state aid to cover the cost of conversion, even though the first farmer had to convert his land at his own cost?

It is clear that there is much consideration required in the establishment of a fair and equitable framework. However, it is worth noting that in the above example the biodiversity underlying the affected Ecosystem Service is likely to be the ultimate victim of all possible scenarios.

This leads us to ask the question: 'how can Ecosystem Services, with their unique profile of service provision with undefined ownership and beneficiaries which are not economically connected to the service, be preserved in the long term?'

As we have discussed, the imposition of thresholds, limits and legislative control with regard to Ecosystem Services is clearly not something which can be rationally applied due to our lack of empirical knowledge. The alternative approach of providing subsidies or incentives to the owners of land or features which provide Ecosystem Services to third party beneficiaries is vulnerable to variation in flow of financing, making Ecosystem Services dependent upon national policy and economic status.

One potential solution to these issues lies in the concept of a truly free market in Ecosystem Services. This approach would develop a movement toward the communal ownership of Ecosystem Service sources by the beneficiaries of these services. Any external influence upon the status quo would require a negotiated process between the perpetrator of any proposed change and the owners/beneficiaries of the Ecosystem Service. This would exclude centralised planning from the equation, and allow direct transactions to take place, which would only proceed if a mutually beneficial arrangement could be made.

This approach would require, in some cases, for the reprioritisation of an ownership hierarchy. Although private ownership of property would be maintained, the ownership of the Ecosystem Service the property provided would be invested in the beneficiaries of the service (which may include the landowner, but not necessarily so). This would clearly require a certain degree of co-operation between the two parties going forward, but both would have a vested interest in an outcome which maintained the service. Damage caused by the landowner would result in his indebtedness to the beneficiaries, and the over-exploitaton of the resource by its beneficiaries would be directly felt by them, leading to responsible use.

Such a concept represents an extension of the idea of property rights (land ownership) to also take into account the rights of those who benefit from the services of the property in those cases where no conscious activity by the owner of the property is required to provide the service.

This approach prevents the situation where a landowner, who is unaware of the service his property provides, is centrally funded to continue doing what has always been done, thereby benefitting purely from ownership. Under these circumstances the beneficiaries of the service become dependent upon the continuation of central funding to a third party landowner and have no control or ability to direct their future use of the resource. This service can be indirectly purchased and destroyed purely by negotiation with the landowner, without any reference to the beneficiaries of the service.

In addition, the provision of central funding may have the perverse effect of making the service more difficult to access by the community at large, as its presence now represents an asset to a landowner, who is likely to wish to 'protect' it for purely personal financial reasons.

This brings us to a somewhat different perspective of the idea of property rights purely as land ownership. If this is considered to be the only factor then all centralised approaches, which concentrate upon ownership as the primary 'right' when concerning land (either through the maintenance of private property as a sacrosanct concept, or through state ownership in order to prevent private ownership of land, via public expenditure) fail to address the right of use of natural services.

By looking at the situation from a purely economic perspective it is possible to draw something of a distinction. Ownership and use of land in order to undertake productive activities is a legitimate free-market ambition. However, the ownership of an Ecosystem Service is not legitimate, when the benefits of ownership is based upon the property rights approach. The landowner has by definition, not expended any productive activity in the generation of an Ecosystem Service. Therefore this coincidental ownership of the land from which the service emanates should not result in any economic benefit. The economic benefits should be assigned to those who hold the rights to use the service. Although the services provided are considered to be 'free', effort and human activity is required to harvest or otherwise retrieve the benefits of such services (particularly those within the category of provisioning services). Therefore the service is not actually free, it costs an expenditure of time and effort to realise its benefits. Of course, such a right of use would also come with responsibilities for the maintenance of the service, which would naturally be in the users best interests.

If it is possible to accept that in the case of Ecosystem Services, property rights (ownership) and service rights (use) are two sides of the same coin, then it is necessary to address the question of ownership with regard to the underlying fundamental of all services, which is biodiversity.

The rather nebulous view that 'biodiversity belongs to everyone' does not actually take us any further in this process. Rather it can easily lead to the concept that no one can be held responsible for it, and even more significantly, there is no benefit to gaining a deeper understanding of it—it just 'is' an amorphous communally owned body in a pseudo-mystical state of being. Under such circumstances our concern about biodiversity is dependent upon personal conscience rather than the reality of the situation (of which we are remarkably uninformed).

As we have previously demonstrated, it is possible to define biodiversity in terms of a commodity. This approach allows for the idea that biodiversity (or more precisely knowledge of biodiversity) as an untapped source of wealth in the actual rather than the abstract sense. To progress this concept further it is necessary to consider the nature of the relationship between commodities and money.

Recognising Sound Money

In the past, commodities have been used by many cultures as a form of currency (i.e. as a medium of exchange which is universally accepted within the culture), such commodity money gains its value from the material of the money itself. Examples of such commodity money include grain, livestock and salt. The problem with such commodity money is that it is literally consumed and requires continuous replenishment to drive the economy.

A solution to this issue comes in the use of precious metals (gold and silver). Although these are generally referred to as commodities, their value does not lie in their utility, but more in their scarcity, desirability and durability. Because these materials are not consumed, but rather stored and universally accepted in trade across borders and between cultures, their use became widespread.

It is of particular interest to note that commodity money can be classified according to a system which bears an uncanny resemblance to the classification for Ecosystem Services. Some forms of commodity money are considered to have direct utility value (grain, livestock etc.), this is paralleled by the provisioning services category of Ecosystem Services. Other forms (such as precious metals) have an aesthetic value, with alternative uses in jewellery and may also represent an indication of social status, such aspects are considered within the realm of cultural services when Ecosystem Services are considered.

The use of commodity money is generally considered to be a somewhat primitive, and rather inconvenient way of undertaking trade. For the direct use of commodities in exchange to operate on a large scale (outside of a local community) requires the materials involved to have a number of characteristics, such as portability, divisibility and longevity which are not always evident in a local commodity currency.

For this reason, commodity currencies experienced limited utility when trade became more universal in nature, with the exception of precious metal coinage, where the value of the coin was in fact held within the metal it was composed of. Even precious metal coinage could cause problems however, either through individuals fraudulently removing a proportion of the coin, or through state sponsored debasement of the metal within the currency.

These issues with commodity money led to the development of an abstract version of commodity money, which became commodity-backed money. This is essentially a promissory note, which allows the holder to redeem it for a fixed amount of the underlying commodity. The most widespread examples of such a scheme is the gold standard, where the currency in circulation is symbolic of an underlying store of precious metal. The vulnerability of such a system to expansion of the money in circulation via fractional reserve banking has been discussed previously.

In more recent times, the value of money is based upon government decree (fiat), there is no underlying commodity which the owner of the currency has a right to access. Arguably, once the USA finally stopped redemption of the US dollar for gold in 1971 (advertised as a temporary measure, but still in force today), the US dollar was in effect backed by oil, as all trade in oil was undertaken in the US currency. However, this is no longer the case, recent political moves have seen oil traded internationally in Chinese Yuan [4].

The departure from any kind of standard against which money is measured (i.e. the Gold Standard) and its replacement with a purely fiat monetary system, although offering greater freedoms for the economic running of a nation, does not allow for any kind of fixed valuation to be made using fiat currencies.

In terms of Ecosystem Services this is particularly problematic, as the value of such services is pegged against what is essentially a fluid benchmark. Over a relatively short period of time this is less of an issue, but over a longer timescale it becomes increasingly relevant, as mainstream economics requires an ongoing devaluation of currency (via inflation) the value of a service becomes increasingly reduced.

This can be seen when one considers the purchasing power of the US dollar over time, since it broke from the gold standard (see Fig. 2 in chapter "Valuing Ecosystems"). If one contends that the provision of a recognised service has a certain value at a given point in time then ongoing economic growth and the necessary inflation which ensues (in the theories of mainstream economics) then one reaches a point where the monetary value of the Ecosystem Service declines to the point where its loss or replacement is no longer a significant economic barrier.

This is true, even assuming the curtailing of the current abilities of businesses and governments obtaining loans via central banking issue of fiat currency, which under the present system is perceived in itself to be economic growth.

One potential method to avoid the systematic devaluation of Ecosystem Services is to consider the biodiversity which underlies Ecosystem Services to be value in and of itself (i.e. a financialised commodity) rather than imposing an external and separate valuation, based upon a financial system which is debt-based and of a purely fiat nature.

Clearly the use of biodiversity as money is not practical when considered as a form of commodity money, issues of ownership and transfer would not make it viable. However, as we have already noted the abstraction of the concept towards commodity backed money means that a biodiversity backed currency would be a realistic idea. This is particularly true when one considers the most recent developments in communication and information technology, which would enable biodiversity to be used as a global financial vehicle.

Applying the Concept

The recent development of distributed ledger technology (exemplified by the popular blockchain protocol) is most notably associated with the rise of cryptocurrencies as a means of creating a decentralised monetary system which is out of the control of central banking and government influence.

However, the basic concept is a means of addressing the question of how individuals can find a way of achieving full consensus over a given problem (in the case of cryptocurrencies the issue is finding a universally accepted means of transactional payment), and furthermore act in unison once an agreement has been found.

If we frame the question in terms of biodiversity, that is 'how can we achieve a universal agreement regarding the value of biodiversity, and act in unison to respect that value?' then we can see how such a technology may provide some kind of steps forward.

A distributed ledger is essentially a decentralised database which is simultaneously accessible to all users. When changes are made to the database, the alterations are updated across all of the terminals accessing it. There is no single, centralised hub controlling these activities.

To date, the most popular use of such technology has been in terms of the production of cryptocurrencies, which have drawn a great deal of attention [5]. Most of these cryptocurrency concepts are based upon the idea that a currency is 'mined' and then released into the blockchain ledger.

Cryptocurrency mining involves the use of computing power (and energy) to find solutions to algorithmic problems, a successful solution is then converted into a unit of cryptocurrency which is then released into the system. Thus, the value of such currencies derives purely from the solution of abstract mathematical problems, using energy and resources with no actual productive benefit to mankind other than the production of a unit of currency. Furthermore, much of the crypto-currency boom is more concerned with the speculative accumulation of units of virtual currency in the hope that they will increase in value (denominated in fiat currencies) rather than their use as a means of exchange in profitable productive activities.

The use of such a decentralised technology to monetise the gaining of biodiversity knowledge through the application of human skill and labour would be a practical means of enabling a biodiversity-backed currency to play a part in the financial system. Under this mechanism, taxonomists would be the miners of this rich seam of knowledge, eco-logical expertise would be the refining process which would extract the value from the raw information. This system would place biologists at the centre of a financialised system which could fully account for the value of biodiversity and ecosystems.

Taxonomic studies do not require extensive capital expenditure, the tools required are simple and the techniques straightforward, if approached using a classical methodology. This is because no taxo-nomic information ever becomes obsolete, but is rather added to by subsequent, more sophisticated methodologies [6]. The main resource used in such studies is human patience, observational abilities and more patience. So human labour would be the input to extract value from the natural world in a straightforward and uncomplicated manner.

Elements of biodiversity would be identified and, once independently verified (to prevent fraudulent manufacturing of biodiversity data) would be entered into the ledger. The information would include morphological and geographical data as a unique digital string, which would be instantaneously accessible to all. In addition to merely logging the data, it would be necessary to identify the level of uniqueness which the record possessed. If there was no similar species within the database, it would receive greater weight than one which was present globally. Such interrogation and analysis would represent the first step in the monetisation of the data within the ledger.

This raises the question of the value to be assigned to each record, how this would operate and who would benefit.

Currently, global sovereign debt stands at $63 trillion (Institute of International Finance 2017). This debt only represents the borrowings of governments, primarily in pursuit of growth stimulation, following mainstream economic theories. This gives an indication of the level of repayment required through either currency debasement (through inflation) or growth (through development and productive activity, making use of resources) which would return the global system to a point where we have actually earned the standard of living that we enjoy today. If we are serious about ecosystems and biodiversity as a key element in our sustainable future, then it is reasonable to equate full knowledge of biodiversity with the current level of debt. It provides a means of side stepping the issue of ignorance of biodiversity being a prerequisite to allow environmental damage in the name of economic progress. In this scenario it is the acquisition of biodiversity knowledge which would incrementally reduce the level of sovereign debt, thereby not only reducing the need to undertake development for growth, but would act as a direct incentive to preserve biodiversity, as it would represent, in tangible economic terms, a nation's wealth.

If we divide the total level of global national debt by the estimated 10 million species on the planet, then we would have a figure of each species being 'worth' approximately $6.3 million of national debt. Of course, this represents an average figure, and ubiquitous species would be valued lower, and uniquely endemic species would be a much more valuable resource.

The system would work through the establishment of a mechanism which would permit offsetting of national debt against biodiversity wealth. One possible means of implementing this would be through the integration of biodiversity wealth into the international Special Drawing Right (SDR) system.

The SDR is an international reserve asset, created by the IMF in 1969 to supplement its member countries' official reserves. It serves as the unit of account of the IMF and some other international organizations. Currently the value of the SDR is based on a basket of five currencies—the U.S. dollar, the euro, the Chinese renminbi, the Japanese yen, and the British pound sterling [7].

The SDR is neither a currency nor a claim on the IMF. Rather, it is a potential claim on the freely usable currencies of IMF members. SDRs can be exchanged for these currencies. Originally the SDR had a commodity base, it was defined as equivalent to 0.888671 grams of fine gold—which, at the time, was also equivalent to one U.S. dollar. Therefore it is not inconceivable that the use of a biodiversity backed currency could be incorporated into the basket of currencies which form the SDR. The next review of the method of valuation of the SDR will take place by September 30, 2021, unless an earlier review is warranted by developments in the interim.

In its current form, the SDR represents an only partial solution to the problem of a universal monetary system. As it currently stands it is an aggregate of currencies which are all individually fiat in nature. There is currently no commodity or other backing to the concept of an SDR. An essay in 2009 by the then Governor of the People's Bank of China [8] addresses the question of the kind of international reserve currency that is needed to secure global financial stability and facilitate world economic growth. The essay notes that theoretically, an international reserve currency should first be anchored to a stable benchmark and issued according to a clear set of rules, therefore to ensure orderly supply; second, its supply should be flexible enough to allow timely adjustment according to the changing demand; third, such adjustments should be disconnected from economic conditions and sovereign interests of any single country. The acceptance of credit-based national currencies as major international reserve currencies, as is the case in the current system, is a rare special case in history.

The use of a biodiversity backed currency offers to meet a number of these requirements, it would offer a stable benchmark and it would be suitably disconnected from prevailing economic conditions. In turn, the use of a biodiversity backed currency system would stop the current view that biodiversity needs to be sacrificed in order to permit economic growth. In fact, its adoption would mean that biodiversity preservation (and its ongoing understanding through knowledge acquisition) would be a key element in the preservation of wealth.

The essay continues by noting that 'the desirable goal of reforming the international monetary system, therefore, is to create an international reserve currency that is disconnected from individual nations and is able to remain stable in the long run, thus removing the inherent deficiencies caused by using credit-based national currencies'.

In this respect the role of the SDR should be strengthened, as it has the features and potential to act as a super-sovereign reserve currency. The current weakness of the SDR is its purely fiat basis, which means that its ability to provide stability is dictated by the economic policies of those nations who have their currencies included within the SDR basket. By introducing a biodiversity backed currency into its basket of currencies, it would be possible to provide a truly stable system. With the environmentally positive side effect that protection of the natural world would become one of the main requirements, as it would be part of the financial foundation of the economic system.

The above discussion has presented an outline by which the use of a biodiversity backed currency as part of the international financial system would not only open the door to a massive increase in the rate by which we gain knowledge of the world around us, but also serve to place biodiversity in the position of being a genuine source of wealth, which would naturally be safeguarded in the same way that any other monetary resource is protected. The next chapter will discuss how such a system would affect a range of stakeholders.

References

1. Gastineau, G. L. (2010). *The Exchange-Traded Funds Manual*. Hoboken: Wiley Finance Series.
2. Mises, L. von. (1949). *Human Action*. New Haven: Yale University Press.
3. TEEB. (2009). *The Economics of Ecosystems and Biodiversity for National and International Policy Makers.*
4. Evans, D. (2018). Shanghai Shakes Up Global Oil Trading. *Nikkei Asian Review.* https://asia.nikkei.com.
5. Jamali, R., Li, S., & Pantoja, R. (2017). Cryptocurrency: Digital Asset Class of the Future—Bitcoin vs Ethereum? *The Economist/Kraken Bitcoin Exchange.*
6. Heywood, V. H. (1979). *Plant Taxonomy.* Institute of Biology: Studies in Biology, No. 5. London: Arnold.
7. *IMF Special Drawing Right Factsheet.* https://www.imf.org/en/About/Factsheets/Sheets/2016/08/01/14/51/Special-Drawing-Right-SDR.
8. Xiaochuan, Z. (2009). Reform the International Monetary System. *BIS Review,* 41.

Effects and Applications

This book has presented the means by which we can alter the way in which we view biodiversity, converting it from a backdrop to human activity and growth into a commodity, the development of knowledge regarding this commodity has a tangible wealth-creating effect.

A biodiversity-backed currency would operate primarily as a debt-relief vehicle via the internationally recognised SDR. Therefore, a nation's wealth would be enhanced through debt relief, rather than through the disturbing concept of a nation 'spending' its biodiversity assets. A currency within the SDR, but only recognised through international consent, would be inaccessible outside of this forum. Therefore, it would be immune from speculative market activity.

We can now look at how this alternative perspective would affect the way in which economic activity is undertaken at various levels, and how different sectors are likely to need to adapt to such a change in worldview.

© The Author(s) 2019
S. Muddiman, *Ecosystem Services*,
Palgrave Studies in Natural Resource Management,
https://doi.org/10.1007/978-3-030-13819-6_6

The Environmental Sector

One of the key roles of the environmental sector under the present paradigm is the framing of legislation designed to protect habitats and species which are considered to be of importance, and subsequently to determine that projects are in compliance with the legislative framework.

One of the main tools in undertaking compliance studies is the Impact Assessment process. This is a procedure undertaken across a broad range of topics. It is carried out according to relevant planning and legislative guidelines which vary according to the prevailing legislation, the subject area and the proposals being considered.

At the most fundamental level, the Impact Assessment process involves the following stages:

- The establishment of a baseline. This involves the gathering together of all the information available about the environmental topic under consideration, within the area of study. Once the data has been gathered, organised and evaluated the overall importance of the area of study is determined. This evaluation may take the form of considering its importance in a geographical scale, or through analysis of its component parts. The type of evaluation undertaken is dependent upon the nature of the legislative framework the assessment is being carried out to conform to. It should be noted that the information which is assessed is generally only relevant to the planning and policy requirements of the nation or state. No legislation can require the acquisition of additional knowledge of biodiversity which is outside of existing corpus of information. Therefore, it is highly unusual for such studies to uncover truly new information. Throughout the process of assessing the effects for projects affecting the environment, the unknown elements of biodiversity remain hidden from view.
- The next stage is to consider the proposed project and determine how it will affect the environmental baseline. This is essentially a descriptive process, considering the various effects and how they may affect the various components of the environment which have been identified. Again, these identified components relate primarily to those

elements which are considered within the relevant planning and legislative framework. These are generally considered to be the features 'of significance' and could be particular habitats, species and more recently Ecosystem Services. Essentially the consideration of what is significant and what is not is driven by policy, rather than ecological theory.

- Once the effects on these significant receptors have been determined, further analysis is carried out to determine to what extent these effects are deleterious to the receptor concerned. This is again based upon existing information and inference, rather than collection of additional data. The result of this procedure is the production of a list of 'significant effects' on those environmental features which policy and legislative processes have determined to be important.
- The normal end of the process is the presentation of means by which any significant adverse effects may be avoided, reduced or compensated for. The aim of the process is to achieve the situation where 'no significant adverse effects' of the proposal remain. This is generally sufficient for policy makers to permit a project to proceed (at least as far as environmental considerations are concerned).

It should be noted that the proposed methodologies for the reduction and compensation of adverse impacts tend to be generic in nature, based upon 'best practice' and generally considered to be effective.

However, the actual checking and confirmation of such measures tends to be only agreed following approval of the project. Of course, if the measures proposed are not effective, then there is usually no recourse as the environmental damage has already occurred. Furthermore, there is no clear and accessible means of collating the results of monitoring a variety of projects, and much of the 'best practice' adopted is not quantifiably effective other than in unique 'case study' scenarios. Failed impact mitigation measures tend not to be reported as openly, as they do not represent good publicity for either the regulators or the practitioners involved.

So how would this process be altered under the alternative system proposed in this book?

In those cases where the financing of a project is not derived from previously profitable activities additional financial input would be required. Under these circumstances the type of assessment described above would be required earlier in the process. Such an assessment would be needed not to achieve approval for the project, but to confirm that funding for the project would be appropriate. This introduces the interests of the project lender into the assessment process, as lending would only be permitted if no environmental damage could be proven. As a result of this, the scope of such assessments would be likely to alter, the focus would be more upon the likelihood of unrecorded biodiversity being present, and to what extent lending would pose risks to the lender through either inadequate baseline information or potentially flawed mitigation proposals. Concerns which focus narrowly on policy-driven protected habitats and species would be less relevant than overall ecological knowledge held within the affected environment.

As a result of 'biodiversity mining' activities, more data would be available and a more comprehensive analysis would be required. The abstract concept of some aspects of the environment being significant and others not would be less important than the accounting process which would determine the value of the area affected through the calculation of a direct monetary value.

In cases where financing from previously profitable activities are available to fund a project, then the responsibility for permitting the works to be undertaken would again fall upon the national, state or regional legislature. The primary focus, under these circumstances would be to ensure that there was no loss of biodiversity assets which could not be counteracted through the benefits of the project in terms of increased employment, taxation or other flows of finance into the public purse. Therefore, an assessment would be required for these purposes, and would be closely scrutinised to ensure that national wealth was not being destroyed for private profit.

Furthermore, any risk components in terms of impact mitigation would need to be much more closely assessed, as speculative measures would require a much higher degree of confidence for their success, due to the direct financial impact of failed measures.

In this paradigm, it is clear that environmental assessments would still play a major role, although the precise skill sets required by practitioners would be likely to alter. This is no different to the current situation, where changes in legislation require attention to be paid to different species, habitats and Ecosystem Services according to the prevailing system. However, the assessments undertaken, in circumstances where biodiversity is considered a commodity used to back financial wealth, would be more wide-ranging in terms of the application of general ecological principles, and also more analytical in output. Concepts of 'significance' would be replaced by cost benefit calculations using market rates. Data analysis would also become a much more important element in assessment procedures.

Outside of the assessment process, there would be a much greater demand for practitioners with taxonomic skills. These skills would be employed by the 'biodiversity mining' community and would represent a key to developing national wealth.

In addition to the description of biodiversity as a means of creating wealth, a demand would arise for ongoing ecological studies as part of the 'biodiversity-financial' sector. This work would be to both secure existing biodiversity resources and provide an ongoing audit of recognised resources. The mere recognition of biodiversity value would not be sufficient in itself to be a proof of the sustainable maintenance of the resource. Periodic auditing would be required, not unlike the auditing needed for other commodity-based financial assets.

The move towards biodiversity backed currency would also have a significant effect upon the activities of conservation management. The emphasis of environmental work would be upon the quantification and understanding of biodiversity in the full meaning of the term. This would result in less importance being placed upon interventionist activities focussed upon single issue conservation management. The protection of a charismatic 'flagship' species would not necessarily be seen as beneficial to the overall biodiversity of an area. Therefore conservation management would become more limited in its extent, but more balanced in terms of the species and habitats being affected.

We can see, therefore, that the environmental sector would expand in scope, although much of the activity would be related to the recognition of national wealth rather than either acting as a compliance vehicle or in a confrontational conservationist versus developer scenario.

Planning and Government

In the current economic regime, policy and legislation regarding the environment finds itself in a condition of tension. Much of the output of government activity in this area is explicitly concerned with protection measures. This is because the need to protect environmental components reveals that there is perceived to be some force or influence which represents a threat, against which protection measures are designed to counteract.

One of the main difficulties in the framing of such protective legislation is the limited background information available to legislators with regard to assigning appropriate priorities.

Under a system of biodiversity backed wealth recognition, there would be much less of an emphasis on the environment as a vulnerable victim of attempts at its destruction. The environment would itself be recognised as a source of wealth and so would become integrated into the accounting procedures of the nation, state or region of which it was a part. The integration of economic factors and valuations would naturally mean that proposals involving environmental damage for short-term economic growth would not gain traction.

The natural environment would have an inherent fiscal value, which would act as a functional deterrent to actions leading to its damage and degradation. Therefore, policies and legislation relating to environmental protection would be less of a matter of importance. However, there would be likely to be an increased focus on policy measures relating to the reuse and recycling of resources, together with programmes for the maintenance of existing infrastructural systems.

The adoption of a currency component backed by a nation's biodiversity assets would allow this biodiversity wealth to act as collateral against which government 'biodiversity bonds' could be issued. This would provide a commodity backed, and therefore stable, issuance for ongoing income. However, unlike the fiat monetary system, there would be more significant limits to the amount which could be borrowed via this system. In addition, there would be an increased emphasis on environmental protection, in its broadest sense, as this would become linked directly with economic well-being.

This arrangement would offer a more self-sustaining process than the allocation of public finances to safeguard biodiversity against private interests and the public funded activities in response to mainstream economic theory. Inevitably there would be less public money spent, which would be focussed on the acquisition of knowledge, for the benefit of all. This change in spending patterns would lead to a less confrontational approach: 'the environment vs progress' as a polarity would be a redundant concept.

With the acquisition of knowledge directly contributing to national wealth, there would be an increase in national policies designed to promote skills in 'biodiversity mining' in order to reduce dependence on external expertise. Such developments would be a major force in driving a wide range of skills, particularly in those areas with potentially rich seams of biodiversity wealth within their borders.

Business and Finance

It is of critical importance to note that the acceptance of a biodiversity backed currency approach will in no way place restriction on development or growth which is based upon business needs driven by genuine profitable commercial activity, as such enterprises would be able to reinvest such profits into additional facilities and expansions without the requirements of seeking additional funding from lenders.

Similarly, any speculative business which was focussed on activities which did not result in effects on the biodiversity wealth of a nation, state or region would be able to receive funding without restrictions on environmental grounds. This is true of both private and public enterprises.

However, there would be strict limitations on those speculative or indebted businesses which wished to grow themselves out of debt through any expansion or development which negatively impacted upon the biodiversity resource of the areas they were active within.

This change in the business landscape would serve to promote an upsurge in the success of businesses with sound business models based upon productive activity rather than the current trend for debt-based growth and speculative profitable activity at some point in the future.

Due to the high demand for areas without any unknown or financially significant biodiversity resources present, empty industrialised areas would be rapidly reused. Business expansion would impose pressures on land occupiers to be more efficient and provide returns on land use. Inefficient business would not be able to borrow additional funds to expand in the hope that upscaling operations would provide greater overall returns. The onus would be on highly efficient land occupancy rather than expansion into greenspaces which may have significant wealth generating capabilities in their own right.

Within the financial sector there would be a much greater emphasis on looking at the environmental consequences of lending. In addition, the reintroduction of a commodity backed currency into the financial system (as a component part of the SDR) would mean that there were limits to the amount of currency available to be created, in contrast to the prevailing fiat system. This would result in a greater degree of caution and analysis before financing for speculative activities would be committed.

Communities

Rural communities would be less prone to disruption when biodiversity was valued on the economic stage. The reduction in available financing would result in less pressure on rural communities as a result of developments. In contrast, an upsurge of activities relating to the recycling and reuse of existing resources would provide an impetus for urban regeneration and increased employment prospects within existing industrialised zones.

Funding for the maintenance of existing natural landscapes, habitats and Ecosystem Services would be driven by the recognition that local communities are the custodians of the biodiversity resources in their vicinity. It would also allow them a greater voice in decision- and policy-making, as a hierarchical approach, using broad-based policy decisions would not be as applicable under the circumstances where the unknown elements of the environment offer tangible wealth-creating potential. Each situation would, of necessity, have to be treated as a separate and distinct case. This would overload the bureaucratic system of hierarchical decision-making, so individual communities would need to be given the trust and freedom to make local decisions for the benefit of the immediate area.

Externally driven interventions, either to improve the environment or to modify it to 'enhance biodiversity' (in the current understanding of the term) would be more circumspect. There would be more caution demonstrated in such decisions and a tendency to retain the status quo. At least until the biodiversity present within the area of interest had been fully characterised.

Of course, the amount of say a local community has would still be dictated by the prevailing political system. However, the presence of biodiversity wealth has the potential to provide communities with a greater influence in the way their environment is controlled and maintained.

Final Comments

This book demonstrates that the current economic and Ecosystem Services 'received wisdom' are actually different manifestations of the same approach to the way humanity is organised. There are certain elements in both which rely upon a top-down hierarchical view of the way in which policy decisions are made. The amount of control which individuals or communities hold in both systems is limited.

Neither system, as it currently stands can be considered to be natural in its means of operation, both rely on political will and policy-driven decision-making. Rather than address the issues of complexity and the unknown (be it the diversity of the natural world or the dynamics of human action) there is a tendency to take our current level of knowledge and crystallise it into 'the truth' through the addition of numerous layers of complication, by way of providing the appearance of knowledge and certainty in what are actually poorly understood systems.

Despite this, alternative views exist, both in terms of economic theory and how Ecosystem Services can be valued. By considering such alternatives, it seems that, although the systems themselves remain complex, the complications can be resolved through the application of simple logical processes.

The recent development of distributed ledger technology offers a functional tool to enable the biodiversity knowledge gap not only to be closed, but also to enable financial benefit to be realised in areas where unrecognised biodiversity occurs.

The benefits of this system do not require the dismantling of the current financial system. By integrating the concept of a biodiversity backed currency into the global SDR, there would be alterations to the economic system which would promote biodiversity investigations and recording of previously unrecognised species, offer financial benefits (through relief of debt) and give the natural environment a financial value which would serve to automatically safeguard its integrity.

Remember, all currencies are fiat in nature. They are not backed by any tangible asset and only have value by decree. Therefore, there is no reason why a consensus cannot be reached amongst all nations that a global biodiversity backed currency should be included into the basket of currencies which is the SDR. Such a decision would accord with the rhetoric of concern for biodiversity loss and the need to value the environment which is espoused without any serious dissent in the international community.

Once such an agreement was made, a ledger would be established and individual nations would be able to begin to claim their share of the value of the currency, through registering the biodiversity present within their borders.

Although there will be many claims that such a system is impractical, I would suggest that the system, though complex is by no means lacking in practicality. I suspect such objections would be more concerned with political expediency than practical concerns.

Index

© The Editor(s) (if applicable) and The Author(s), under exclusive license   **161**
to Springer Nature Switzerland AG, part of Springer Nature 2019
S. Muddiman, *Ecosystem Services*, Palgrave Studies
in Natural Resource Management, https://doi.org/10.1007/978-3-030-13819-6

The manufacturer's authorised representative in the EU is Springer
Nature Customer Service Centre GmbH, Europaplatz 3, 69115 Heidelberg,
Germany. If you have any concerns regarding our products, please
contact ProductSafety@springernature.com

Printed and bound by CPI Group (UK) Ltd, Croydon, CR0 4YY
05/05/2026
02103216-0002